Plant Systematics and Evolution

Entwicklungsgeschichte und Systematik der Pflanzen

Supplementum 2

Genome and Chromatin: Organization, Evolution, Function

Symposium, Kaiserslautern, October 13-15, 1978

Edited by W. Nagl,
V. Hemleben, and F. Ehrendorfer

Springer-Verlag Wien New York

Prof. Dr. Walter Nagl

Universität Kaiserslautern, Federal Republic of Germany

Prof. Dr. Vera Hemleben

Institut für Biologie II, Universität Tübingen,
Federal Republic of Germany

Prof. Dr. Friedrich Ehrendorfer

Institut für Botanik, Universität Wien, Austria

© 1979 by Springer-Verlag/Wien

Softcover reprint of the hardcover 1st edition 1979

With 101 Figures

Library of Congress Cataloging in Publication Data

Main entry under title:
(Plant systematics and evolution: Supplementum; 2) 1. Plant genetics—Congresses. 2. Chromatin—Congresses. 3. Chromosomes—Congresses. I. Nagl, Walter. II. Hemleben, V.. 1939— III. Ehrendorfer, Friedrich. IV. Series. QH433.G47 581.8'732 79-20416

ISBN-13:978-3-7091-8558-2   e-ISBN-13:978-3-7091-8556-8
DOI: 10.1007/978-3-7091-8556-8

Preface

At a round table discussion on the eukaryotic chromosome sponsored by the *Deutsche Forschungsgemeinschaft* in Düsseldorf, February 1978, the botanists among the participants felt that plant systems were under-represented. In this unsatisfactory situation, Professor V. HEMLEBEN, Tübingen, suggested another meeting to discuss actual problems and results concerning botanical chromosome research. Professor W. NAGL was willing to organize a symposium at the University of Kaiserslautern, and Professor F. EHRENDORFER, Wien, contacted the Springer-Verlag, Vienna – New York, to explore the possibility of publishing the results of this symposium in the form of a supplement volume to the journal *Plant Systematics and Evolution*. The conference took place on 13–15 October 1978 in the Department of Biology of the University of Kaiserslautern and was attended by 40 participants from 11 universities between Hamburg and Vienna.

Emphasis of this Chromosome Symposium was given to three aspects, which do not attract major interest at *large* international congresses: 1. Discussion and Demonstration of technical details which cannot be found in published papers (so-called tricks). 2. Orientation about actual trends and results in our understanding of the organization, evolution, and function of the plant genome at the level of the DNA (gene), the level of chromatin, and the level of the karyotype. 3. Presentation of hypotheses and models which may be stimulating for further research. Moreover, younger students should have the possibility to present their results and to discuss them with more experienced scientists.

The success of this symposium rests on the fact that everyone was keen to unfold his methods, results, and ideas in an unconventional way, and because scientist working in different fields were ready to talk with each other: molecular biologists, cytologists, physiologists, evolutionists etc. An important result of this meeting was the expression of the participants' will to set up inter-universitary groups which will co-work in the fields of cloning of plant genes, and the study of differential DNA replication.

Not all papers given at the Symposium are included in this volume; those presented deal with different aspects such as methodology, actual data, new interpretations of well-known facts, etc. In general, we feel that this volume is representative for what is done in plant genome and chromosome research in Germany and Austria at the moment, what trends can be seen in this field, and what lines of evidence need supplementation.

We are thankful to the President of the University of Kaiserslautern, Professor H. EHRHARDT, who allowed the meeting and supported its organization by the provision of technical staff. We are also grateful to the Freundeskreis der Universität Kaiserslautern for financial support, and the burgomaster of Kaiserslautern, Mr. B. BARTHEL, for a reception in the Kasimir castle. The remarkable efforts of research students and staff members of the Division of Cell Biology are particularly acknowledged. For drawing many of the diagrams shown in this volume we thank Mrs. U. GILLE. Finally, we wish to express our sincere thanks to all contributors and participants for extensive information and discussion.

Kaiserslautern, Tübingen, and Vienna   WALTER NAGL
September 1979   VERA HEMLEBEN
                                          FRIEDRICH EHRENDORFER

Contents

Introductory Lecture

Genome Organization and Evolution

Differential DNA Replication

Gene Numbers and Transcription

Chromatin Structure

Nucleotype and Heterochromatin

Introductory Lecture

Pl. Syst. Evol., Suppl. 2, 3—25 (1979)

Department of Biology, University
of Kaiserslautern, Federal Republic of Germany

Search for the Molecular Basis of
Diversification in Phylogenesis and Ontogenesis

By

Walter Nagl, Kaiserslautern

Key Words: Evolution, speciation, differentiation, morphogenesis, polyploidy, DNA amplification, karyotype repatterning, repetitive DNA.

Abstract: Two basic phenomena of living matter, splitting (speciation etc.)
during evolution and cell differentiation during somatic development, are
only poorly understood. In this article the evidence for the following
suggestions is discussed: 1) The phenomena of phylogenetic and ontogenetic
diversification are independent of respectively the adaptation/selection system
and the differential transcription control system; speciation and cell
differentiation precede adaptation and regulation. 2) The mechanisms which
lead to phylogenetic and ontogenetic diversification may be found among the
following ones: polyploidization, differential DNA replication (*e. g.* saltatory
replication, amplification), karyotype repatterning (including differential
chromatin condensation). The non-coding repetitive DNA sequences may play
an important role in organismic and cellular diversification in changing growth
rate parameters and the regulatory programs which are part of the genome
organization itself.

Phenomena and Problems

One of the most remarkable differences between inanimate and
living matter can be seen in the relative stability of the former, and the
permanent diversification of the latter. There is a slow diversification of
basic programs (*"Baupläne"*) which are rapidly realized by ontogenetic
cell differentiation and morphogenesis in billions of individuals that
compose the living world, and whereby the phylogenetic diversification
is recapitulated in principle ("Haeckel'sches Grundgesetz"). Molecular
biology has shown that the ontogenetic sequence of differentiation
follows the phylogenetic one; this may occur because the more
conservative members of any family of DNA sequences are more
reiterated and transcribe more copies of RNA (FLICKINGER 1975), or
because the DNA of most organisms still contains the genes of all

1*

ancestors (genes of the "housekeeping" functions and of low complexity organisms); in adult organisms only the most recently evolved genes become active, controlling the function of the most recently evolved complex organs (see also the article by WENZEL in this volume). BRITTEN & DAVIDSON (1971) pointed out that there is an obvious ontogenetic recapitulation of phylogeny with respect to regulatory programs, and that the study of differential gene regulation and the study of evolution must be considered two sides of one and the same coin (GALAU & al. 1976). MIKSCHE & HOTTA (1973) concluded that variation in repetitive DNA may be as much part of ontogeny as of phylogeny. So I think that we can learn much about ontogenetic differentiation and gene regulation from understanding phylogenetic speciation, and we may find a way to elucidate speciation, if we know how the diversification of cells is controlled in an embryo. However, both phenomena — the evolution and the function of living matter — must causally be determined in the structural organization of its molecules, cells and organs. A central role in the complex systems of phylogenesis and ontogenesis must be attributed to the chromosomes. Therefore, the common theme, linking the various aspects of this symposium may be the structural organization of the chromosomes as it relates to chromosome and species evolution as well as to chromosome and cell function.

Let me first draw your attention to some points in evolution, which need new consideration, and to whose clarification this symposium may be helpful.

1) The rates of molecular evolution (in terms of nucleotide and amino acid sequence diversification) seem to be virtually independent of rates of organismal (morphological) evolution (WILSON & al. 1977). Molecular evolutionists were slow to recognize this surprising and intriguing fact. For instance, they excepted that morphologically conservative organisms should have experienced slower gene sequence evolution than organisms that had evolved unusually rapidly at the morphological level. They were impressed by the observation that the phylogenetic trees constructed from macromolecular sequences usually resemble those based on morphological evidence, with regard to branching order (e.g. GOODMAN 1976). This congruence in branching order is explained simply by supposing that when a species splits into two reproductively isolated daughter species, divergence between them begins to take place at both the morphological and the sequence level. Congruence, however, does not require that the rate of morphological change be geared to the rate of sequence change (WILSON & al. 1977). In addition, the correlation between the degree of sequence difference, estimated by electrophoretic comparison of many proteins, and the

taxonomic distance between organisms does not necessarily imply a cause-and-effect relationship.

There are good examples of prominent discrepancies in the organismal evolutionary rate and the macromolecular or gene evolution rate. Placental mammals, for instance, have experienced rapid organismal evolution compared with lower vertebrates, of which frogs are a typical example: Although there are thousands of frog species,

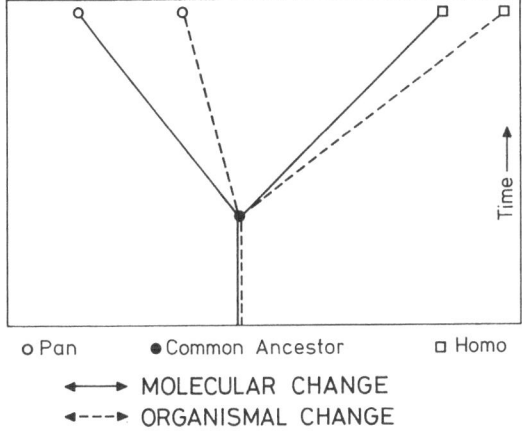

o Pan   ●Common Ancestor   □ Homo

◄——► MOLECULAR CHANGE
◄----► ORGANISMAL CHANGE

Fig. 1. Diagram to illustrate the contrast between biological evolution and molecular evolution since the divergence of the human and the chimpanzee lineages according to KING & WILSON (1975). More biological change has taken place in the human lineage than in the chimpanzee lineage, while protein and nucleic acid sequences indicate that as much change has occurred in chimpanzee genes as in human genes

they are all put together into a single order, *Anura*, whereas the phylogenetically younger placental mammals are divided into at least 16 orders. Clearly, organismal evolution has been slow in frogs relative to mammals; yet, at the molecular level of the gene and gene products, evolution has been just as rapid. Another noteworthy example concerns ape and human evolution. Since humans and chimpanzees had a common ancestor, much more phenotypic change has occured in the human lineage than in that of the chimpanzee, but no such tendency is evident at the sequence level (WILSON 1976; Fig. 1). Species within a genus of mice, frogs, or flies can differ more from each other in gene nucleotide or amino acid sequence than does a human from a chimpanzee, although these two species are classified in different taxonomic families (KING & WILSON 1975, CHERRY & al. 1978). In other

cases, no significant allozymic differences have been found in some
Hawaiian *Drosophila* species despite their morphological divergence
(SENE & CARSON 1977). These findings clearly indicate that morphological evolution and biochemical evolution in structural genes can proceed
at independent rates.

2) When talking about genome, chromatin and karyotype
evolution and function there is always the need for the classification
of the organisms under consideration. The most fundamental set of
similar organisms is the species. In the whole of biology, however, there
is no concept that is so much discussed, and yet so little understood, as
both the definition and the origin of species (MAYR 1963, WHITE 1969,
SHORT 1976). Although natural selection offers a simple and acceptable
way of explaining how the environment exerts a constant pressure to
adapt and improve within a species, the central problem is how do
new species originate? This origin always must preceed selection
pressure and is, therefore, independent of it. SHORT (1976) pointed out
that it is important to realize that terms such as species, sub-species,
race and strain are man's attemps to divide what is essentially a
continuum of genetic change into arbitrary series of discrete sets.
Moreover, there are sufficient examples of both animal and plant
interspecific hybrid fertility to invalidate any simple definition of
species status based on the dictum of hybrid sterility.

The traditional view that most evolutionary change is gradual and
cumulative within lineages has recently been challenged by the
proposition that the majority of evolutionary change is concentrated
and linked to speciation and other splitting events (cladogenesis, or
rectangular evolution: STANLEY 1975, NAGL 1978). It may be that the
origin of most organismal novelty accompanies speciation episodes, and
that other phyletic and *genetic* changes account only for the "fine-
tuning" of the *Baupläne* laid down during speciation in order to adapt
to the environment. The specification process thus would be effectively
decoupled from phyletic evolution, and much of the theory of phyletic
gradualism would be simply irrelevant to important evolutionary
patterns (AVISE 1977).

3) The question on the origin of species throws us back to a more
fundamental question: What is the basic mechanism of evolution, what
is it that gives a direction to evolution? Does evolution proceed
according to the neo-Darwinian ideas by positive selection of the fittest,
or according to the neutral mutation — random drift hypothesis of
KIMURA (1968, 1977), i.e. in a non-Darwinian way? Are the initial steps
in the genetic process of speciation the result of accident rather than
adaptation? (DOVER 1978). DOBZHANSKY (1976) stated that the working
hypothesis which seems to him most fruitful is that species are not

accidents but adaptive devices through which the living world has developed itself to master a progressively greater range of environments and ways of living. Chance, however, in the form of random gene frequency drift, and in the form of neutral and nearneutral mutations, may actually play a larger part than previously considered (KIMURA 1976). Is the obvious direction of evolution given by fitness or complexity? It seems that the bacteria are already the fittest organisms. The increase in complexity may be a consequence of the process by which a self-optimizing system optimizes its organization with respect to a locally defined fitness potential (SAUNDERS & HO 1976). But is evolution an optimizing process? (AUSLANDER & al. 1978). Is the fitness of phenotypes or genotypes subject to selection? (LLOYD 1977). Recent data suggest that natural selection is based more directly on the macromolecules, their secondary structure and higher order structure, and on the transcription and translation stability rather than on ecological conditions (MAZIN 1976, RUSSEL & al. 1976, PIECHOCKI & al. 1977, EIGEN & SCHUSTER 1978 a, b).

Perhaps a synthesis of all those contradicting views is possible with the system theory of RIEDL (1976, 1977). According to this theory, the systemic conditions developed by the organization of the organism operate as selection, as a negative feed-back condition between genotype and phenotype, limiting the ecological adaptation. As RIEDL (1977) pointed out, the master key to unlock the problem of life must fit into all strata, from the molecular to the organismic and population level, simultaneously. When the problem of explaining *Gestalt* in terms of DNA structure should be solved in terms of life as a compromise between two conflicting selective components, infrastructure (organization) versus adaptability, then the key may be found in the regulation of ontogenetic differentiation and phylogenetic diversification. According to several authors this regulation may be brought about by the noncoding repetitive DNA sequences. Then the ontogenetic and phylogenetic variation in the amount and composition of non-coding repetitive DNA fractions could be understood as evolutionary strategy in order to optimize the control DNA, which is necessary to regulate the complex networks of stage- and tissue-specific gene activations and inactivations and to maintain the complex patterns of differential gene activity (NAGL 1978). DNA and chromosomes must not be seen as static constancies, but as dynamic, variable structures (LIMA-DE-FARIA 1975, SHARMA 1977). The ontogenetic changes can be understood as a predefined process which adapts the somatic cells of the individuals to the phylogenetic process of speciation and the Bauplan encoded by the species' genome, i.e. genes and non-coding DNA (NAGL 1977 b, 1978).

Gene Duplication, Polyploidization, and
DNA Diversification

After this rough review of some aspects of organismic versus molecular diversification and evolution, a few genetic and molecular genetic aspects will be discussed.

Gene duplication is well established as one of the most important factors in evolution (OHNO 1970). COHEN & al. (1977) have shown that enzymes belonging to the same biosynthetic pathway could derive from common ancestors. Enzyme polymorphism (isoenzymes) are a good example for gene duplication and diversification, independently from the question of whether this polymorphism has adaptive advantage in a heterogeneous environment (JOHNSON 1976) and is directly subject to selection (CLARKE 1975), or whether this variation is maintained by the stochastic interaction between neutral mutations and genetic drift (ALLENDORF 1978). Also the C-terminal portions of the core histones, H 2a, H 2b, H 3, and H 4 seem to have eveled from a common ancestral protein (REECK & al. 1978). Moreover, an important kind of genetic change is the duplication of part of a gene (internal gene duplication), resulting in an elongated strand of DNA in the genome, which then codes for an elongated protein (BARKER & al. 1978).

Polyploidy is a common evolutionary phenomenon among plants and several groups of animals (STEBBINS 1971, BOGART & TANDY 1976, FERRIS & WHITT 1977). The reason for the enormous succesfulness of polyploidy may be seen in the accompanying higher biochemical versatility (BARBER 1970, ADAMS & ALLARD 1977) or other kinds of adaption (LEWIS & SUDA 1976). In allopolyploid species, which are "fixed heterozygotes", all the individuals express the multiple enzyme phenotypes. This multiplicity of enzymes may extend the range of environments in which normal development can take place (GOTTLIEB 1976). But also in autopolyploids quantitative and qualitative differences in zymograms and protein profiles have been recorded (e.g. NAKAI 1977), as well as differences in the physiology of autopolyploid plants (TAL & GARDI 1976). It is worthwhile to note that diploid-like chromosome pairing in meiosis, and hence reproductive stability, is under genetic control in sexually reproducing polyploids (e.g. JAUHAR 1977, WATANABE 1977).

Another important aspect of polyploidy may be seen in the fact that the multiple structural genes diverge in structure and function subsequent to the origin of polyploids. Thus, new unique patterns of gene expression in different tissues or different levels of activity become possible (for a plant see HART & LANGSTON 1977, for two animals see FERRIS & WHITT 1977 a). Hence, the process of speciation may be manifested in the diploidization of autotetraploid species, or even caused by it (Paunović 1977, Ferris & Whitt 1977 b).

Gymnosperms and angiosperms have evidently followed different evolutionary strategies. Elaborating on suggestions of Miksche & Hotta (1973), Ehrendorfer (1976) and Grant (1976) speculated that large chromosomes and the greater quantity and

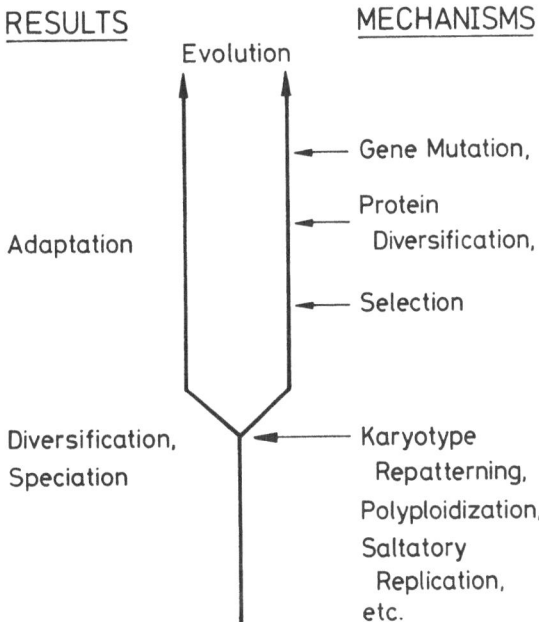

Fig. 2. Diagram to illustrate the effect of some of the suggested mechanisms which are leading to cladogenesis (speciation) and orthogenesis (adaptation of lineages)

redundancy of DNA have prevented polyploidy and more active divergent evolution in gymnosperms. In angiosperms, on the other hand, small chromosomes and smaller quantities of less repetitive DNA seem to have brought about more cytogenetic and evolutionary versality and better avenues for cyclic diversification on diploid and polyploid levels of organization.

Although this brief discussion of gene and genome duplication and diversification indicates some selective advantage of these kinds of mutation, it cannot be deduced from this section that speciation is a consequence of adaption. Recent analyses rather indicate that the speciation process, i.e. the basic step of phylogenetic diversification, occurs prior to the acquisition of distinct adaptations and is largely fortuitous (GOTTLIEB 1976, NAGL 1978; Fig. 2).

When genetic differentiation is a function of time, unrelated to the number of cladogenetic events as shown by AVISE (1978), then the question again arises: What molecular event can be related to speciation, phylogenetic diversification in general? While genome multiplication by germ-line polyploidization is evidently one way to explain the process of cladogenesis, selective increase of certain sequences, called either "secondary DNA" (HINEGARDNER 1976), or "nonfunctional DNA" (SCHMIDTKE & al. 1978), or simply "non-coding repetitive DNA" (BRITTEN & DAVIDSON 1971, NAGL 1978) may be another way, more often realized during evolution than polyploidization.

Non-Coding Repetitive DNA

A balanced growth (and diminution, respectively) of chromosomes during evolution involving numerous minute interstitial additions (or deletions) of DNA has been repeatedly described (LIMA-DE-FARIA 1975, SUMNER & BUCKLAND 1976, MACGREGOR 1978). The cytological findings are in good agreement with biochemical results on an increase of all reassociation fractions when the 2 C value is enhanced in the course of macroevolution (BALDARI & AMALDI 1976; see also GHARROTT & al. 1977). Related species differ mainly in the amount of rpetitive DNA (reviewed by NAGL 1978). Those unique sequences, the proportion of which varies in addition to the reiterated sequences, very likely do not code for proteins or stable RNAs: In sea-urchins, for instance, only 10 % of the unique sequences seem to represent coding genes (GALAU & al. 1976). The molecular mechanism of interstitial DNA increase is probably unequal crossing over (for a model see SMITH 1976, PERELSON & BELL 1977), or saltatory replication and insertion of the extra copies of DNA.

A second phenomenon is the origin of clusters of new repetitive sequences in chromosome regions that appear heterochromatic. For instance, heterochromatin was identified to be responsible for the variation in nuclear DNA contents in squirrels (differences of 70 %! MASCARELLO & MAZRIMAS 1977). A substantial fraction of this excess DNA consists of sequences that band as satellites in neutral CsCl or Cs_2SO_4-Ag^+ density gradients. Saltatory changes in constitutive heterochromatin played also an important role in the karyotype evolution of Australian ants (IMAI & al. 1977). Estimates of the 2 C DNA content made by BENNETT & al. (1977) in 16 taxa of the genus *Secale* ranged from 14.5 pg to 17.8 pg, and Giemsa C-banding patterns showed considerable interspecific and intraspecific variation and several instances of polymorphism for large telomeric C-bands. This strongly supports the notion that the major evolutionary change in chromosome structure in *Secale* involved the addition of heterochromatin by saltatory amplification at, or close to, the telomeres. Similar situations have been recorded in the genus *Anacyclus* (SCHWEIZER & EHRENDORFER 1976) and *Scilla* (GREILHUBER & SPETA 1978), and in Australian grasshoppers (MIKLOS & NANKIVELL 1976). Heterochromatin also plays a key role in the basic biology of *Uromys caudimaculatus*, an endemic Australian murid rodent (BAVERSTOCK & al. 1976). While northern chromosomal races display substantial distal Giemsa C+ blocks, none of the southern individuals has fixed distal heterochromatin, but their karyotype contains from six to nine mitotically stable supernumerary chromosomes all of which are totally Giemsa C-positive.

The function of the repetitive DNA in both euchromatin and heterochromatin is not yet fully understood, but several lines of evidence indicate biological significance. MIKLOS & NANKIVELL (1976) are sure that heterochromatin is implicated in some form of alteration in the meiotic recombination system, and that satellite DNA function in regulating the level and position of recombination. Also EDSTRÖM (1976) suggested that a large fraction of the nuclear DNA, variable in amount between different species, has an exclusive meiotic function and therewith a direct evolutionary role.

Another aspect of addition (and deletion) of repetitive DNA is the nucleotypic effect of nuclear DNA, which is independent of its information content and base sequence (BENNETT 1973, 1976). The nuclear DNA content has been found to be positively correlated with DNA synthesis time and minimum cell cycle time, the duration of meiosis, minimum generation time and so on. Heterochromatin could exert a nucleotypic effect which is different from that of euchromatin (NAGL 1974). Perhaps amplified non-coding DNA is effective as a coordinator of different growth parameters and gene activities.

Thus the nucleotypic effect of DNA might just be a special case of some regulatory function of the non-coding, repetitive sequences of the genome (for some general speculations see BRITTEN & DAVIDSON 1971, ZUCKERKANDL 1976, DAVIDSON & al. 1977). The suggestion that the interspersed reiterated sequences have a control function to those genes which they are adjacent to, is particularly suggestitive in the light of somatic amplification of repetitive DNAs during cell differentiation (see below). On the other hand, the clustered repetitive sequences which are localized in the heterochromatin, may be the main breaking points of chromosomes thus being the substrate for karyotype repatterning (JONES & al. 1972) and increase in the amount of repetitive DNA (BAIMAI

1975, 1977). Actually, heterochromatin which was neglected by geneticists for long time because of its genetic inertness, now becomes the key structure in karyotype evolution, which in turn seems to play a key role in organismic diversification (e.g. King 1977).

Before considering karyotype repatterning, the fate of repetitive sequences during further evolution has to be taken into consideration. There is now good evidence for both, dispersal of repetitive DNA sequences throughout the genome (Bendich 1977) and divergence of repetitive sequences (Flavell & al. 1977). Harpold & Craig (1977) report a rate of approximately 0.1 % sequence divergence per million years in the repetitive DNA sequences of sea urchins, while non-repetitive DNA sequences were found to diverge at a rate of approximately 0.22 % per million years (Harpold & Craig 1978). These results indicate that repetitive DNA sequences are strongly conserved during evolution, although highly repetitive sequences (satellite DNA) and spacer DNA may diverge more rapidly than do the moderately repetitive, interspersed sequences. These results are consistent with the hypotheses that highly repetitive DNAs function independent of their base sequence in controlling karyotype structure (Lima-de-Faria 1975), while the moderately repetitive DNAs may act as regulatory elements (Britten & Davidson 1971). We are, however, far away from understanding the conservation or divergence of repetitive sequences, as no generalization of the above statement is yet possible (see also Mizuno & al. 1976, Mazrimas & Hatch 1977, Arnheim & Southern 1977). Jones (1976) suggested that the variation in repetitive DNA is advantageous to sexual organisms in avoiding the complications of both polyploidy and B chromosome systems, two other ways of nuclear DNA variation.

Karyotype Repatterning

The immense variety of karyotypes found in extant species is unmistakable evidence that the process of evolution is associated with karyotypic change. The question whether the chromosome changes are a cause or a consequence of speciation has been debated intensely for many years, and as evolution evidently operates along different lines in different groups of organisms (e.g. quantitative DNA changes, or karyotype repatterning) there has been no unequivocal answer, as is often the case with biological problems (Fredga 1977). Many evolutionists, however, believe today that changes in the macro-structure of the genome are more important than point mutations in facilitating the rapid phases of evolution and particularly of speciation (White 1969, Turleau & al. 1972, Ayala 1975, Bush 1975, Hatch & al.

1976, BUNCH & al. 1976, SHORT 1976, WILSON 1976, BAVERSTOCK & al. 1977 a, b, BEDO 1977, BUNCH & al. 1977, HENEEN 1977, SINGH & RÖBBELEN 1977). Robertsonian fusion and fission seem to be the main mechanisms in the evolution of karyotypes. FINAZ & al. (1977) studied the evolution of chromosome 1 in the primates as it may have occurred during the last fifty-million years and reached the conclusion that this chromosome has maintained its banding pattern and has always carried the same genes, but undergone some structural rearrangements. This permanency of structural genes emphasizes that the difference between related species must be searched for in the organization and regulation of these genes rather than in their mutation.

LEVIN & WILSON (1976) found a close relationship between the evolution of seed plants and the evolution of their karyotypes i.e. increasing chromosome number with complexity — a rule that has also been realized by STEBBINS (1950, 1971, 1976). In general, the evolution of new chromosome numbers and new species is much more intensive in herbs than in shrubs and trees of Angiosperms, it is very low in Conifers, and zero in Cycads. JONES & BROWN (1976) analysed 22 species of *Crepis* with 2 C DNA values between 4.72 and 40.25 pg (the polyploid species show values up to 46.97 pg). The advanced annual species had less DNA, but more chromosomes with median centromeres than the primitive, perennial species, and a higher karyotype symmetry. The opposite strategy has, however, been found too (STEBBINS 1958, SARBHOY 1977). Such results indicate that structural changes of the karyotype are not random accidents but directional rearrangements (e.g. MEZGER-FREED 1977, IMAI & MARUYAMA 1978). The direction of karyotype repatterning is evidently a consequence of a hierarchy in the genome which allows evolution only along the narrow lines preestablished by its rigid molecular configuration (LIMA-DE-FARIA 1975, 1976 a, b, HATCH & al. 1976). LIMA-DE-FARIA (*loc. cit.*) defined the "chromosome field" as the dynamic system of molecular interactions between the various chromosome regions in which all parts are interrelated and in equilibrium so that a change in any one chromosome segment will affect the whole. That means that the basic mechanism of chromosome evolution resides in the physical properties and steric configurations of its constituent molecules. The hierarchy in the control of chromosome function may be organized as follows:

1) At the basic level we find the unique genes coding for specified proteins.

2) At the second level we may see the repeated genes, which show a high output of gene products; moreover, these genes can control somewhat the efficiency of the unique genes, as they produce rRNA and tRNAs.

3) The third level is occupied by the moderately repeated DNA sequences, which are interspersed with DNA of level one and two, and which are involved in the regulation of gene activity.

4) At the top level there is highly repetitive DNA, which is variable in quantity, located in massive tandem arrays, organized in such a way that it determines the macrostructure of the chromosomes and karyotypes, and which controls the evolution of the genome, and thus the evolution of new species.

The final question then is the following: How can non-coding DNA, karyotype repatterning etc. be involved in the control of speciation? Very likely because changes in the amount and position of non-coding DNA within the genome result in changes in gene regulation. And this suggestion may explain the discrepancies between organismic and genetic evolution which have been discussed above. The differences which characterize species when they are first formed may be numerous and largely of a regulatory nature, and not due to gene mutations (CARSON 1976). Moreover, adaptation may take place by changes not in the structural gene loci coding for certain enzymes, but by regulatory changes affecting the amount of gene products (McDONALD & al. 1977). Such a regulatory locus adjacent to the structural gene has been investigated for the xanthine dehydrogenase locus of Drosophila (CHOVNIK & al. 1976, EDWARDS & al. 1977). A simple translocation between two yeast chromosomes was shown to cause overproduction of iso-2-cytochrome, evidently because an abnormal controlling region became associated with the structural gene (SHERMAN & HELMS 1978). The arrangement of heterochromatin in Drosophila chromosomes seems also to affect the structure and integration of the ribosomal genes (HARFORD & ZUCHOWSKI 1977).

In this connection the control role of jumping genes, insertion elements, extrachromosomal elements, paramutations etc. should be considered. However, these poorly understood aspects of genome organization will not be discussed here (for examples see e.g. FINCHAM & SASTRY 1974, CULLIS 1977, HAGEMANN & BERG 1977, MINAMORI & al. 1977).

We have now arrived at the conclusion that the non-coding DNA sequences are more important in the control karyotype evolution and gene regulation, and thus in the control of phylogenetic diversification, than are mutations of the genes proper. If the statement which was given at the beginning of this review, is correct namely that studying evolution and speciation and studying somatogenesis and differentiation are two sides of the same coin, then we must find similar phenomena and mechanisms in both cladogenesis and cell differentiation. Let us, therefore, take a look now at ontogenetic diversification and its control.

Cell Differentiation

The diversification of cells during somatogenesis is one of the most striking phenomena in nature and has attracted cell biologists for long time. The most widely accepted relevant theory is that of genome constancy in all cells, and differential gene activation in order to achieve cell and tissue specific differentiation. Although many steps of the control mechanisms of gene activation and gene expression have been elucidated (such as the role of non-histone chromosomal proteins, RNA polymerase species, post-transcriptional and translational regulation etc.), the basic question is still unanswered: What factor controls the realization of species-specific morphology and physiology during ontogenesis in spite of all the individual variability?

Anatomical studies led HAECKEL (1866) more then one century ago to formulate his *Grundgesetz*, and molecular biology confirmed his ideas recently (FLICKINGER 1975, GALAU & al. 1976). Since we have considered polyploidy, karyotype repatterning, and changes in the non-coding repetitive DNA sequences as basic factors in phylogenetic diversification of living matter, we now ask whether we can find similar phenomena in ontogenesis and whether they are linked to cell differentiation.

Actually, we may see the same mechanisms effective in ontogenetic and phylogenetic variation of genome organisation and nuclear DNA content (Fig. 3; NAGL 1978): Somatic polyploidy is the most common change in nuclear DNA content and chromosome number during development of many plants and animals (GEITLER 1953, D'AMATO 1974, NAGL 1976, 1978). Karyotype reconstruction has been observed in a tissue-specific manner in the rainbow trout (OHNO & al. 1965), and in other cases tissue-specific organization and condensation of the chromatin may lead to the same result as actual numerical or internal karyotype reconstruction (for further interesting ideas see PANDEY 1977, ZIEG & al. 1977). Characteristic chromosomal changes are also often involved in malignancy (e.g. GERMAN 1974, HECHT & KAISER-McCAW 1977, ATKIN & BAKER 1977, BLOCH-SHTACHER & SACHS 1977, LEVAN & al. 1977, ZANKL & ZANG 1978). Although only a parallelism of chromosomal reconstructions or aneuploid and carcinogenesis is proven, but not a causal relationship, the former appear to be a *conditio sine qua non* for the latter. The suggestion that karyotype structure plays a role in differentiation and mis-differentiation is consistent with the hypothesis of some hierarchy within the genome (e.g. LIMA-DE-FARIA 1975).

Saltatory replication (amplification) of genes is well known in ribosomal DNA. This DNA is evidently under separate replication control in both oocytes and somatic cells, so that the number of ribosomal genes varies considerably between individuals of the same species, and even between tissues of the same individual (see tables in NAGL 1975, 1977 a, 1978, BUIATTI 1977). Occasionally, also structural genes may be amplified (e.g. in DNA puffs of giant chromosomes:

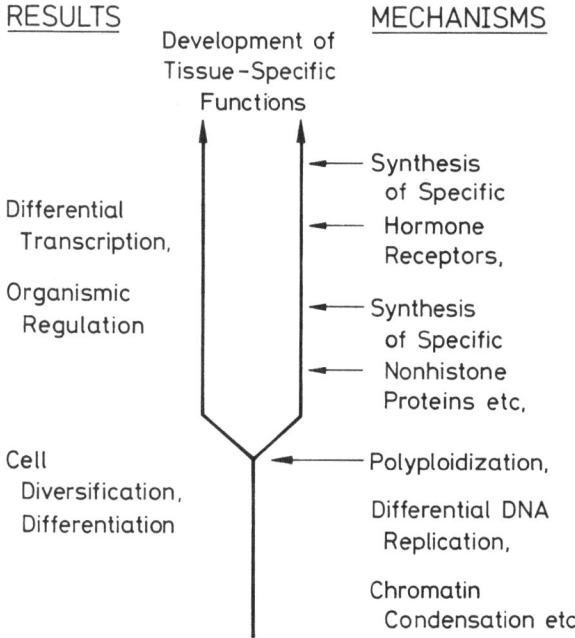

Fig. 3. Diagram to illustrate the effect of some somatic mechanisms which are leading to cell differentiation and specialized function of the differentiated cells. Notice the similarity to the phylogenetic events shown in Fig. 2

WINTER & al. 1977, or the dihydrofolate reductase genes in methotrexate-resistant variants of cultured murine cells: ALT & al. 1978). However, non-coding repetitive DNA sequences are apparently more often amplified than genes (see tables in NAGL, *loc. cit.*). STROM & DORFMAN (1976) and STROM & al. (1978) reported the amplification of repetitive DNA sequences during chicken cartilage and retina differentiation (for the retina cells of the Guinea pig see SCHMALENBERGER & NAGL 1979) KESSLER & RECHES (1977) and SCHÄFFNER & NAGL (1979) found an increase in repetitive DNA during phase change in the ivy, and NAGL & al. (1979) during floral induction of the elder.

Variation of nuclear DNA content during somatogenesis due to differential DNA replication (amplification, under-replication, metabolic DNA — for discussion of these variants see NAGL 1976, 1978, SCHEUERMANN 1978) may be part of cell differentiation and ontogenesis, as are gain and loss of DNA during speciation and phylogenesis (MIKSCHE & HOTTA 1973, GALAU & al. 1976, NAGL 1978), because both ontogenetic and phylogenetic diversification are subject to the same control mechanisms.

Only little is known about how repetitive DNA sequences can regulate gene activity. There are only different types of speculative models available so far. BRITTEN & DAVIDSON (1969, 1971) and DAVIDSON & al. (1977) suggest that the mass of repetitive DNA transcribes a regulatory RNA (rather than a regulatory protein), which coordinates the activation of gene batteries. GUILLÉ & QUETIER (1973) suggested that the repetitive DNA species may serve as qualitative and quantitative "command units" for transcription, and ZUCKERKANDL (1976) interpreted the "excess" DNA in eukaryotes as an impediment to transcription of coding genes. According to this idea, the length of the barrier DNA would influence the rate of transcription at two structural levels. But the repetitive DNA sequences may also comprise recognition sites, as suggested by the eukaryotic operon model (GEORGIEV & al. 1974). Finally the non-coding DNA may control growth processes through its nucleotypic effects on cell cycle duration, cell size, and so on (BENNETT 1973; several aspects and models are compared by NAGL 1976). Whatever model may be more realistic, sufficient data are now available to support the hypothesis of a central role of repetitive DNA sequences in the control of gene activity, cell differentiation, and morphogenesis. The questions, whether and how, extra DNA copies are inserted, excised and transposed in the genome, and where they are actually inserted (adjacent to the gene, or as "intron" within the gene?) will not be discussed in this introductory lecture, but it should be kept in mind that an increasing body of data becomes available in relation to this aspect, too.

There is a last aspect, which will be briefly touched on. Both phylogenesis and ontogenesis, the evolution of more complex species and the realization of the speciesspecific *morphé*, show some kind of direction. Where does this come from? In the past this difficult question was either ignored, or answered in philosophical or transcendental terms. MONOD (1970) pointed out that chance and necessity may be the motor of evolution, and RIEDL (1976, 1977) has shown that macroevolution can only be understood as a feedback system between the adaptation to the environment and the systemic conditions brought about by the organization of the organism. I think that the variation of repetitive DNA may act as a switch in the systemic conditions during evolution and morphogenesis. The changes in the genome, which are caused by chance, can only be realized, if they are within the limit of the systemic conditions (controlled by the hierarchy of the society, the organism, the organ, the tissue, the cell type, the organelle, the macromolecule), so that a *a posteriori* direction becomes visible. However, it cannot be excluded that there is also a *a priori* direction, which is controlled by thermodynamic properties of the DNA itself, and which directs the two basic characteristics of living systems: the capability of self-instruction and the capability of self-organization (EIGEN & SCHUSTER 1978 a, b, PRIGOGINE 1978; see also EISENBERG 1978).

The idea of a central role of differential replication of non-coding, repetitive DNA sequences in ontogenetic and phylogenetic diversification is just a hypothesis (the so-called "DNA optimization model": NAGL 1977 c, 1978), but a hypothesis that is scientifically acceptable: It is logic, because (1) it is consistent with many data, which cannot be explained by other theories, (2) it is experimentally examinable, (3) it falsifies existing theories (the dogma of DNA constancy), and (4) it is falsificable by further research. POPPER (1968) has shown that the progress of science consists of continually attempting to falsify existing theories. The principal objective of a scientist is not to be right, but to contribute to the promotion of science (LØVTRUP 1977). In this sense I hope that the *Kaiserslautern Chromosome Symposium* may contribute to the progress of genome research at different levels.

Support by the *Deutsche Forschungsgemeinschaft* (grant Na-107/3) is gratefully acknowledged.

References

ADAMS, W. T., ALLARD, R. W., 1977: Effect of polyploidy on phosphoglucose isomerase diversity in *Festuca microstachys*. — Proc. Nat. Acad. Sci. (U.S.) **74**, 1652—1656.

ALLENDORF, F. W. 1978: Protein polymorphism and the rate of loss of duplicate gene expression. — Nature **272**, 76—78.

ALT, F. W., KELLEMS, R. E., BERTINO, J. R., SCHIMKE, R. T., 1978: Selective multiplication of dihydrofolate reductase genes in methotrexate-resistant variants of cultured murine cells. — J. Biol. Chem. **253**, 1357—1370.

ARNHEIM, N., SOUTHERN, E. M., 1977: Heterogeneity of the ribosomal genes in mice and men. — Cell **11**, 363—370.

ATKIN, N. B., BAKER, M. C., 1977: Abnormal chromosomes and number 1 heterochromatin variants revelaed in C-banded preparations from 13 bladder carcinomas. — Cytobios **18**, 101—109.

AUSLANDER, D., GUCKENHEIMER, J., OSTER, G., 1978: Random evolutionary stable strategies. — Theor. Popul. Biol. **13**, 276—293.

AVISE, J. C., 1977: Is evolution gradual or rectangular? Evidence from living fishes. — Proc. Nat. Acad. Sci. (U.S.) **74**, 5083—5087.
— 1978: Variances and frequency distributions of genetic distance in evolutionary phylads. — Heredity **40**, 225—237.
AYALA, F. J., 1975: Genetic differentiation during the speciation process. — Evol. Biol. **8**, 1—78.
BAIMAI, V., 1975: Heterochromatin and multiple inversions in a *Drosophila* chromosome. — Canad. J. Genet. Cytol. **17**, 15—20.
— 1977: Chromosomal polymorphisms of constitutive heterochromatin and inversions in *Drosophila*. — Genetics **85**, 85—93.
BALDARI, C. T., AMALDI, F., 1976: DNA reassociation kinetics in relation to genome size in four amphibian species. — Chromosoma **59**, 13—22.
BARBER, H. N., 1970: Hybridization and the evolution of plants. — Taxon **19**, 154—160.
BARKER, W. C., KETCHAN, L. K., DAYHOFF, M. O., 1978: A comprehensive examination of protein sequences for evidence of internal gene duplication. — J. Mol. Evol. **10**, 265—281.
BAVERSTOCK, P. R., WATTS, C. H. S., HOGARTH, J. T., 1976: Heterochromatin variation in the Australian rodent *Uromys caudimaculatus*. — Chromosoma **57**, 397—403.
— — 1977 a: Chromosome evolution in Australian rodents. I. The Pseudomyinae, the Hydromyinae and the *Uromys/Melomys* group. — Chromosoma **61**, 95—125.
— — ROBINSON, A. C., ROBINSON, J. F., 1977 b: Chromosome evolution in Australian rodents. II The *Rattus* group. — Chromosoma **61**, 227—241.
BEDO, D. G., 1977: Cytogenetics and evolution of *Simulium ornatipes* SKUSE (*Diptera: Simuliidae*). I. Sibling speciation. — Chromosoma **64**, 37—65.
BENDICH, A. J., 1977: Dispersal of satellite DNA sequences throughout the muskmelon genome and the nature of families of repeated DNA sequences in plants. In: TS'O, P. O. P., (Ed.): Molecular Biology of the Mammalian Genetic Apparatus, pp. 63—77. — Amsterdam: North-Holland.
BENNETT, M. D., 1973: Nuclear characters in plants. — Brookhaven Symp. Biol. **25**, 344—366.
— 1976: DNA amount, latitude, and crop plant distribution. — Environm. Exp. Bot. **16**, 93—108.
— GUSTAFSON, J. P. SMITH, J. B., 1977: Variation in nuclear DNA in the genus *Secale*. — Chromosoma **61**, 149—176.
BLOCH-SHTACHER, N., SACHS, L., 1977: Identification of a chromosome that controls malignancy in Chinese hamster cells. — J. Cell. Physiol. **93**, 205—212.
BOGART, J. P., TANDY, M., 1976: Polyploid amphibians: three more diploidtetraploid species of frogs. — Science **193**, 334—335.
BRITTEN, R. J., DAVIDSON, E. H., 1971: Repetitive and non-repetitive DNA sequences and a speculation on the origins of evolutionary novelty. — Quart. Rev. Biol. **46**, 111—138.
BUIATTI, M., 1977: Gene amplification and tissue cultures. In REINERT, J., BAJAJ, S. (Eds.): Plant Cell, Tissue and Organ Culture, pp. 358—374. — Berlin: Springer.
BUNCH, T. D., FOOTE, W. C., SPILLETT, J. J., 1976: Translocations of acrocentric chromosomes and their implications in the evolution of sheep (*Ovis*). — Cytogen. Cell Genet. **17**, 122—136.

Bush, G. L., 1975: Modes of animal speciation. — Ann. Rev. Ecol. Syst. **6**, 336—362.
— Case, S. M., Wilson, A. C., Patton, J. L., 1977: Rapid speciation and chromosomal evolution in mammals. — Proc. Nat. Acad. Sci. (U.S.) **74**, 3942—3946.
Carson, H. L., 1976: The unit of genetic change in adaptation and speciation. — Ann. Missouri Bot. Gard. **63**, 210—223.
Cherry, L. M., Case, S. M., Wilson, A. C., 1978: Frog perspective on the morphological divergence between humans and chimpanzees. — Science **200**, 209—211.
Chovnick, A., Gelbart, W., McCarron, M., Osmond, B., Candido, E. P. M., Baillie, D. L., 1976: Organization of the *rosy* locus in *Drosophila melanogaster*: Evidence for a control element adjacent to the xanthine dehydrogenase structural element. — Genetics **84**, 233—255.
Clarke, B., 1975: The contribution of ecological genetics to evolutionary theory: detecting the direct effects of natural selection on particular polymorphic loci. — Genetics **79**, 101—113.
Cohen, G. N., Saint-Girons, I., Truffa-Bachi, P., 1977: The evolution of biosynthetic pathways. — Tibs **4**, 97—99.
Cullis, C. A., 1977: Molecular aspects of the environmental induction of heritable changes in flax. — Heredity **38**, 129—154.
D'Amato, F., 1964: Endopolyploidy as a factor in plant tissue development. — Caryologia **17**, 41—52.
Davidson, E. H., Klein, W. H., Britten, R. J., 1977: Sequence organization in animal DNA and a speculation on hnRNA as a coordinate regulatory transcript. — Devel. Biol. **55**, 69—84.
Dobzhansky, Th., 1976: Organismic and molecular aspects of species formation. In Ayala, J. F. (Ed.): Molecular Evolution, pp. 95—105. — Sunderland, Mass.: Sinauer.
Dover, G., 1978: DNA conservation and speciation: adaptive or accidental? — Nature **272**, 123—124.
Edström, J.-E., 1976: Meiotic versus somatic transcription with special reference to Diptera. In Dahlem Workshops on Organization and Expression of Chromosomes, pp. 301—316. — Berlin: Dahlem Workshops.
Edwards, T. C. R., Candido, E. P. M., Chovnick, A., 1977: Xanthine dehydrogenase from *Drosophila melanogaster*. A comparison of the kinetic parameters of the pure enzyme from two wild-type isolalleles differing at a putative regulatory site. — Mol. Gen. Genet. **154**, 1—6.
Ehrendorfer, F., 1976: Evolutionary significance of chromosomal differentiation patterns in gymnosperms and primitive angiosperms. In Beck, C. B. (Ed.): Origin and Early Evolution of Angiosperms, pp. 220—240. — New York: Columbia Univ. Press.
Eigen, M., Schuster, P., 1978 a: The hypercycle. A principle of natural self-organization. B. The abstract hypercycle. — Naturwissenschaften **65**, 7—41.
— — 1978 b: — C. The realistic hypercycle. — Naturwissenschaften **65**, 341—369.
Eisenberg, H., 1978: Chemistry-thermodynamics of biological systems. — Tibs **3**, N 4—N 5.
Ferris, S. D., Whitt, G. S., 1977 a: Duplicate gene expression in diploid and tetraploid loaches (*Cypriniformes, Cobitidae*). — Biochem. Genet. **15**, 1097—1112.

FERRIS, S. D., WHITT, G. S., 1977 b: Loss of duplicate gene expression after polypliodization. — Nature **265**, 258—260.

FINAZ, C., VAN CONG, N., COCHET, C., FREZAL, J., DE GROUCHY, J., 1977: Fifty-million-year evolution of chromosome 1 in the primates. Evidence from banding and gene mapping. — Cytogen. Cell Genet. **18**, 160—164.

FINCHAM, J. R. S., SASTRY, G. R. K. 1974: Controlling elements in maize. — Ann. Rev. Genet. **8**, 15—50.

FLAVELL, R. B., RIMPAU, J., SMITH, D. B., 1977: Repeated sequence DNA relationships in four cereal genomes. — Chromosoma **63**, 205—222.

FLICKINGER, R., 1975: Relation of an evolutionary mechanism to differentiation. — Differentiation **3**, 155—159.

FREDGA, K., 1977: Chromosomal changes in vertebrate evolution. — Proc. R. Soc. (London) **B 199**, 377—397.

GALAU, G. A., CHAMBERLIN, M. E., HOUGH, B. R., BRITTEN, R. J., DAVIDSON, E. H., 1976: Evolution of repetitive and non-repetitive DNA. In AYALA, J. F. (Ed.): Molecular Evolution, pp. 200—224. — Sunderland, Mass.: Sinauer.

GEITLER, L., 1953: Endopolyploidie und endomitotische Polyploidisierung. — Wien: Springer.

GEORGIEV, G. P., VARSHAVSKY, A. J., RYSKOV, A. P., CHURCH, R. B., 1974: On the structural organization of the transcriptional unit in animal chromosomes. — Cold Spring Harbor Symp. Quant. Biol. **38**, 869—884.

GERMAN, J., 1974: Chromosomes and Cancer. — New York: Wiley.

GOODMAN, M., 1976: Protein sequences in phylogeny. In AYALA, J. F. (Ed.): Molecular Evolution, pp. 141—159. — Sunderland, Mass.: Sinauer.

GOTTLIEB, L. D., 1976: Biochemical consequences of speciation in plants. In AYALA, J. F. (Ed.): Molecular Evolution, pp. 123—140. — Sunderland, Mass.: Sinauer.

GRANT, W. F., 1976: The evolution of karyotype and polyploidy in arboreal plants. — Taxon **25**, 75—84.

GREILHUBER, J., SPETA, F., 1978: Quantitative analyses of C-banded karyotypes, and systematics in the cultivated species of the *Scilla sibirica* group (*Liliaceae*). — Pl. Syst. Evol. **129**, 63—109.

GUILLÉ, E., QUETIER, F., 1973: Heterochromatic, redundant and metabolic DNA's: a new hypothesis about their structure and function. — Progr. Biophys. Mol. Biol. **27**, 123—142.

HAECKEL, E., 1866: Generelle Morphologie der Organismen. Allgemeine Grundzüge der organischen Formwissenschaft, mechanisch begründet durch die von CH. DARWIN reformierte Deszendenztheorie. — Berlin: Reimer.

HAGEMANN, R., BERG, W., 1977: Vergleichende Analyse der Paramutationssysteme bei höheren Pflanzen. — Biol. Zentralbl. **96**, 257—301.

HARFORD, A. G., ZUCHOWSKI, C. I., 1977: The effect of X chromosome heterozygosity on the structure of the ribosomal genes in *Drosophila melanogaster*. — Cell **11**, 389—394.

HARPOLD, M. M., CRAIG, S. P., 1977: The evolution of repetitive DNA sequences in sea urchins. — Nucl. Ac. Res. 4, 4425—4437.

— — 1978: The evolution of nonrepetitive DNA in sea urchins. — Differentiation **10** 7—11.

HART, G. E. LANGSTON, P. J., 1977: Chromosomal location and evolution of isozyme structural genes in hexaploid wheat. — Heredity **39**, 263—277.

HATCH, F. T., BODNER, A. J., MAZRIMAS, J. A., MOORE, D. H. II., 1976: Satellite DNA and cytogenetic evolution. DNA quantity, satDNA and karyotypic variations in kangaroo rats (genus *Dipodomys*). — Chromosoma **58**, 155—168.

HECHT, F., KAISER-McCAW, B., 1977: Chromosomes and genes in human cancer cells: multidisciplinary approaches to a unitary genodemographic hypothesis. — Chrom. Today **6**, 357—361.

HENEEN, W. K., 1977: Chromosomal polymorphism in isolated populations of *Elymus (Agropyron)* in the Aegean. IV. General discussion. — Hereditas **86**, 237—244.

HINEGARDNER, R., 1976: Evolution of genome size. In AYALA, J. F. (Ed.): Molecular Evolution, pp. 179—199. — Sunderland, Mass.: Sinauer.

IMAI, H. T., MARUYAMA, T., 1978: Karyotype evolution by pericentric inversion as a stochastic process. — J. Theor. Biol. **70**, 253—261.

— CROZIER, R. H., TAYLOR, R. W., 1977: Karyotype evolution in Australian ants. — Chromosoma **59**, 341—393.

JAUHAR, P. P., 1977: Genetic regulation of diploid-like chromosome pairing in *Avena*. — Theor. Appl. Genet. **49**, 287—295.

JOHNSON, G. B., 1976: Enzyme polymorphism and adaption in Alpine butterflies. — Ann. Missouri Bot. Gard. **63**, 248—261.

JONES, K. W., PROSSER, J., CORNEO, G., GINELLI, E., BOBROW, M., 1972: Satellite DNA constitutive heterochromatin, and human evolution. In PFEIFFER, R. A. (Ed.): Modern Aspects of Cytogenetics: Constitutive Heterochromatin in Man, pp. 45—61. — Stuttgart: Schattauer.

JONES, R. N., 1976: Genome organization in higher plants. — Chrom. Today **5**, 117—130.

— BROWN, L. M., 1976: Chromosome evolution and DNA variation in *Crepis*. — Heredity **36**, 91—104.

KESSLER, B., RECHES, S., 1977: Structural and functional changes of chromosomal DNA during aging and phase change in plants. — Chrom. Today **6**, 237—246.

KIMURA, M., 1976: Population genetics and molecular evolution. —Johns Hopkins Med. J. **138**, 253—261.

— 1977: Preponderance of synonymous changes as evidence for the neutral theory of evolution. — Nature **267**, 257—276.

— 1978: Evolutionary rate at the molecular level. — Nature **217**, 624—626.

KING, M. C., 1977: The applications of molecular evolution to systematics: rates, regulation, and the role of natural selection. — Ann. Missouri Bot. Gard. **64**, 181—183.

— WILSON, A. C., 1975: Evolution at two levels. Molecular similarities and biological differences between humans and chimpanzees. — Science **188**, 107—116.

LEVAN, A., LEVAN, G., MITTELMAN, F., 1977: Chromosomes and cancer. — Hereditas **86**, 15—30.

LEVIN, D. A., WILSON, A. C., 1976: Rates of evolution in seed plants: Net increase in diversity of chromosome numbers and species numbers through time. — Proc. Nat. Acad. Sci. (U.S.) **73**, 2086—2090.

LEWIS, W. H., SUDA, Y., 1976: Diploids and polyploids from a single species population: temporal adaption. — J. Heredity **67**, 391—393.

LIMA-DE-FARIA, A., 1975: The relation between chromomeres, replicons, operons, transcription units, genes, viruses and palindromes. — Hereditas **81**, 249—284.

— 1976 a: The chromosome field. I. Prediction of the location of ribosomal cistrons. — Hereditas **83**, 1—22.

LIMA-DE-FARIA, A., 1976b: The chromosome field. II. The location of "knobs" in relation to telomeres. — Hereditas **83**, 23—34.
— 1976c: The chromosome field. IV. The distribution of chromomere gradients in relation to kinetochore and telomeres. — Hereditas **84**, 19—34.
LLOYD, D. G., 1977: Genetic and phenotypic models of natural selection. — J. Theor. Biol. **69**, 543—560.
LØVTRUP, S., 1977: The Phylogeny of Vertebrata. — London: J. Wiley.
MACGREGOR, H. C., 1978: Some trends in the evolution of very large chromosomes. — Phil. Transact. R. Soc. (London), **B 283**, 309—318.
MASCARELLO, J. T., MAZRIMAS, J. A., 1977: Chromosomes of antelope squirrels (genus *Ammospermophilus*): a systematic banding analysis of four species with unusual constitutive heterochromatin. — Chromosoma **64**, 207—217.
MAYR, E., 1963: Animal Species and Evolution. — Cambridge, Mass.: Harvard Univ. Press.
MAZIN, A. L., 1976: Evolution of DNA structure: direction, mechanism, rate. — J. Mol. Evol. **8**, 211—249.
MAZRIMAS, J. A., HATCH, F. T., 1977: Similarity of satellite DNA properties in the order *Rodentia*. — Nucl. Ac. Res. **4**, 3215—3227.
McDONALD, J. F., CHAMBERS, G. K. DAVID, J., AYALA, F. J., 1977: Adaptive response due to changes in gene regulation: a study with *Drosophila*. — Proc. Nat. Acad. Sci. (U.S.) **74**, 4562—4566.
MEZGER-FREED, L., 1977: Chromosomal evolution in a haploid frog cell line: implications for the origin of karyotype variants. — Chromosoma **62**, 1—15.
MIKLOS, G. L. G., NANKIVELL, R. N., 1976: Telomeric satellite DNA functions in regulating recombination. — Chromosoma **56**, 143—167.
MIKSCHE, J. P. HOTTA, Y., 1973: DNA base composition and repetitive DNA in several conifers. — Chromosoma **41**, 29—36.
MINAMORI, S., INOUE, Y., ITO, K., SHIMIZU, A., 1977: Extrachromosomal element delta in *Drosophila melanogaster*. X. Enhancement in female and male recombinations. — Jap. J. Genet. **52**, 87—93.
MIZUNO, S., ANDREWS, C., MACGREGOR, H. C., 1976: Interspecific "common" repetitive DNA sequences in salamanders of the genus *Plethodon*. — Chromosoma **58**, 1—31.
MONOD, J., 1970: Le Hasard et la Nécessité. — Paris: Ed. du Seuil.
NAGL, W., 1974: The role of heterochromatin in the control of cell cycle duration. — Nature **249**, 53—54.
— 1975: Organization and replication of the eukaryotic chromosome. — Progr. Bot. **37**, 186—210.
— 1976: Zellkern und Zellzyklen. — Stuttgart: Ulmer.
— 1977a: Organization and replication of the eukaryotic chromosome. — Progr. Bot. **39**, 132—152.
— 1977b: The evolution of chromosomal DNA redundancy: ontogenetic lateral, versus phylogenetic tandem changes. — Nucleus **20**, 10—27.
— 1977c: The DNA optimization model for speciation and cytodifferentiation. — Chrom. Today **6**, 151—152.
— 1978: Endopolyploidy and polyteny in differentiation and evolutions. Towards an understanding of nuclear DNA variation during ontogeny and phylogeny. — Amsterdam: North-Holland.
— FRISCH, B., FRÖLICH, E., 1979: Extra-DNA during floral induction. — Pl. Syst. Evol., Suppl. **2**, 111—118.
NAKAI, Y., 1977: Variations in esterase isozymes and some soluble proteins in

diploids and their induced autotetraploids in plants. — Jap. J. Genet. **52**, 171—181.

OHNO, S., 1970: Evolution by gene duplication. New York: Springer.

— STENIUS, C., FAISSI, E., ZENZES, M. T., 1965: Post-zygotic chromosomal rearrangements in rainbow trout (*Salmo irideus* GIBBONS). —Cytogenetics **4**, 117—129.

PANDEY, K. K., 1977: Generation of multiple genetic specifities: origin of genetic polymorphism through gene regulation. — Theor. Appl. Genet. **49**, 85—93.

PAUNOVIĆ, D., 1977: A cytogenetic analysis of the genus *Neodendrocoleum* (*Triclada, Paludicola*) from lake Ohrid. — Chromosoma **63**, 161—180.

PERELSON, A. S., BELL, G. I., 1977: Mathematical models for the evolution of multigene families by unequal crossing over. — Nature **265**, 304—310.

PIECHOCKI, R., BERG, W., BERGMANN, A., 1977: Das Ausmaß asymmetrischer Basenverteilung in eukaryotischen Genen. — Theor. Appl. Genet. **49**, 265—271.

POPPER, K. R., 1968: The Logic of Scientific Discovery (2nd ed.). — London: Hutchinson.

PRIGOGINE, I., 1978: Time, structure and fluctuations. — Science **201**, 777—785.

REEK, G. R., SWANSON, E., TELLER, D. C., 1978: The evolution of histones. — J. Mol. Evol. **10**, 309—317.

RIEDL, R., 1976: Strategie der Genesis. — München: Piper.

— 1977: A systems-analytical approach to macro-evolutionary phenomena. — Quart. Rev. Biol. **52**, 351—370.

RUSSEL, G. J., WALKER, P. M. B., ELTON, R. A., SUBAK-SHARPE, J. H., 1976. Doublet frequency analysis of fractionated vertebrate nuclear DNA. — J. Mol. Biol. **108**, 1—23.

SARBHOY, R. K., 1977: Cytogenetical studies in the genus *Phaseolus* LINN. III. Evolution in the genus *Phaseolus*. — Cytologia **42**, 401—413.

SAUNDERS, P. T., HO, M. W., 1976: On the increase in complexity in evolution. — J. Theor. Biol. **63**, 375—384.

SCHÄFFNER, K.-H., NAGL, W., 1979: Differential DNA replication involved in transition from juvenile to adult phase in *Hedera helix*. — Pl. Syst. Evol., Suppl. **2**, 105—110.

SCHEUERMANN W., 1978: Überprüfung der cytologischen Phänomene bei *Vicia faba*, die zu PELC'S Hypothese einer "metabolischen" DNA beitrugen. — Cytobiologie **17**, 232—245.

SCHMALENBERGER, B., NAGL, W., 1979: Different DNA content, chromatin condensation, and transcription activity in retina cell nuclei of the Guinea-pig. — Pl. Syst. Evol., Suppl. **2**, 119—125.

SCHMIDTKE, J., KUNZ, B., ENGEL, W., 1978: Gene numbers and genetic variability in *Ciona intestinalis* and *Branchiostoma lanceolatum*. — Canad. J. Genet. Cytol. **20**, 61—70.

SCHWEIZER, D., EHRENDORFER, F., 1976: Giemsa banded karyotypes, systematics, and evolution in *Anacyclus* (*Asteraceae, Anthemideae*). — Pl. Syst. Evol. **126**, 107—148.

SENE, F. M., CARSON, H. L., 1977: Genetic variation in Hawaiian *Drosophila*. IV. Allozymic similarity between *D. silvestris* and *D. heteroneura* from the islands of Hawaii. — Genetics **86**, 187—198.

SHARMA, A. K., 1977: Evidence of dynamism as a basis of chromosomal control of genetic reactions. — Nucleus **20**, 4—10.

SHERMAN, F., HELMS, C., 1978: A chromosomal translocation causing overproduction of iso-2-cytochrome c in yeast. — Genetics **88**, 689—707.

SHORT, R. V., 1976: The origin of species. In AUSTIN, C. R., SHORT, R. V. (Eds.) The Evolution of Reproduction, pp. 110—148. — Cambridge: Univ. Press.

SINGH, R. J., RÖBBELEN, G., 1977: Identification by Giemsa technique of the translocations separating cultivated rye from three wild species of *Secale*. — Chromosoma **59**, 217—225.

SMITH, G. P., 1976: Evolution of repeated DNA sequences by unequal crossover. — Science **191**, 528—535.

STANLEY, S. M., 1975: A theory of evolution above the species level. — Proc. Nat. Acad. Sci. (U.S.) **72**, 646—650.

STEBBINS, G. L., 1950: Variation and Evolution in Plants. — New York: Columbia University Press.

— 1958: Longevity, habitat and release of genetic variability in the higher plants. — Cold Spring Harbor Symp. Quant. Biol. **23**, 365—377.

— 1966: Chromosome variation and evolution. — Science **152**, 1463—1469.

— 1971: Chromosomal Evolution in Higher Plants. — London: Arnold.

— 1976: Chromosomes, DNA and plant evolution. — Evol. Biol. **9**, 1—34.

STROM, C. M., DORFMAN, A., 1976: Amplification of moderately repetitive DNA sequences during chick cartilage differentiation. — Proc. Nat. Acad. Sci. (U.S.) **73**, 3428—3432.

— MOSCONA, M., DORFMAN, A., 1978: Amplification of DNA sequences during chicken cartilage and neural retina differentiation. — Proc. Nat. Acad. Sci. (U.S.) **75**, 4451—4454.

SUMNER, A. T., BUCKLAND, R. A., 1976: Relative DNA contents of somatic nuclei of ox, sheep, and goat. — Chromosoma **57**, 171—175.

TAL, M., GARDI, I., 1976: Physiology of polyploid plants: water balance in autotetraploid and diploid tomato under low and high salinity. — Physiol. Plant **38**, 257—261.

TURLEAU, C., DE GROUCHY, J., KLEIN, M., 1972: Phylogenie chromosomique de l'homme et des primates hominiens (*Pan troglodytes, Gorilla gorilla* et *Pongo pygmaeus*). Essai de reconstitution du caryotype de l'ancêstre commun. — Ann. Génét. **15**, 225—240.

WATANABE, K., 1977: The control of diploid-like meiosis in polyploid taxa of *Chrysanthemum* (*Compositae*). — Jap. J. Genet. **52**, 125—131.

WHITE, M. J. D., 1969: Chromosomal rearrangements and speciation in animals. — Ann. Rev. Genet. **3**, 75—98.

WILSON, A. C., 1976: Gene regulation in evolution. In AYALA, F. J. (Ed.): Molecular evolution, pp. 225—234. — Sunderland, Mass.: Sinauer.

— CARLSON, S. S., WHITE, T. J., 1977: Biochemical evolution. — Ann. Rev. Biochem. **46**, 573—639.

WINTER, C. E., DE BIANCHI, A. G., TERRA, W. R., LARA, F. J. S., 1977: The giant DNA puffs of *Rhynchosciara americana* code for polypeptides of the salivary gland secretion. — J. Insect Physiol. **23**, 1455—1459.

ZANKL, H., ZANG, K. D., 1978: Chromosomenanomalien und Tumorentstehung. — Klin. Wochenschr. **56**, 7—16.

ZIEG, J., SILVERMAN, M., HILMEN, M., SIMON, M., 1977: Recombinational switch for gene expression. — Science **196**, 170—172.

ZUCKERKANDL, E., 1976: Gene control in eukaryotes and the C-value paradox. "Excess" DNA as an impedient to transcription of coding sequences. — J. Mol. Biol. **9**, 73—104.

Address of the author: Prof. Dr. WALTER NAGL, Department of Biology, The University, P. O. Box 3049, D-6750 Kaiserslautern, Federal Republic of Germany.

Genome Organization and Evolution

Pl. Syst. Evol., Suppl. 2, 29—40 (1979)

Institute of Biology II, Department of Genetics,
The University of Tübingen, German Federal Republic

DNA Reassociation Studies and Considerations on the Genome Organization and Evolution of Higher Plants

By

Wolfgang Wenzel and **Vera Hemleben**, Tübingen

Key Words: Angiosperms, *Brassicaceae*, *Matthiola incana*, *Brassica pekinensis*. — DNA reassociation, genome organization, interspersion pattern.

Abstract: The genomes of the crucifers *Matthiola incana* and *Brassica pekinensis* were studied by DNA/DNA reassociation to compare reiteration frequencies and genome organization. Approximately 20% of the *M. incana* DNA is made up of highly repetitive sequences (about 1,000,000 copies per 1 C-genome) which cannot be found in the genome of *B. pekinensis*. For an explanation, a hypothesis on the phylogenesis of crucifers is proposed. The palindromes of *M. incana* DNA are arranged in regularly spaced clusters. This is a common organization in eukaryotes. A short-period interspersion (*Xenopus*-like) pattern of repetitive DNA elements is predominant in the *M. incana* genome. This too seems to be a common pattern among eukaryotes. If we assume that the elements of such a pattern are genes which are active only during embryogenesis (HAECKEL's principle in molecular terms), a possible explanation of the C-value paradox in terms of "gene-excess" becomes available.

Since BRITTEN & KOHNE (1968) we know about the existence of repetitive DNA sequences within the eukaryotic genome. A number of theories have been put forward to explain the function and evolution of these sequences (for a review see NAGL 1976). However, the enormous variation in nuclear DNA contents (C value paradox) is not yet fully understood. Investigations of plant genomes may help to discover some correlation between DNA sequences organization and evolutionary events, in view of the fact that higher plants cannot escape environmental factors like animals which are much more mobile. In response to a changing environment, plants have evolved a number of special genetic features, such as polyploidy. In addition to genome mutations of this kind, fundamental changes in the arrangement of DNA sequences linked in covalent fashion, brought about during the evolutionary process, are evidently necessary. Plants might, therefore, be much better candidates for the elucidation of DNA sequence organization, genome function, and genome evolution than animals. Hence, the study of the DNA sequence arrangement may be of great interest comparing different plant species, particularly of cultivated plants.

Material and Methods

DNA Purification and Isolation. DNA was isolated from seedlings of *Matthiola incana* (L.) R. Br., strain 17, and *Brassica pekinensis* (Lour.) Ruprecht cv. "Granat", grown under sterile conditions, by means of a phenol extraction procedure. The DNA was either sonicated (Branson ultrasonifier), in order to be sheared to various fragment lengths, or was treated in a Virtis homogenizer 60 (according to Davidson & al. 1973). Then the DNA was passed over HAP columns, using standard conditions as given by Britten & al. (1974), in order to get rid of non-DNA components (elution with 50 mM NaPB), RNA contamination, and single-stranded DNA (eluted with 80 mM NaPB). Native DNA was collected in fractions of 400 mM NaPB. Aliquots of this DNA were tested for purity by melting (T_m-value; reduction in T_m-value depending upon DNA fragment length; the hyperchromicity was 30% for *M. incana*, strain 17, DNA and 36% for *B. pekinensis* DNA) and reassociation (Gilford spectrophotometer 210), before the DNA was subjected to further experiments. *E. coli* DNA (strain B) was treated in the same way as described for plant DNAs. The average fragment size of the DNA was determined by agarose gel electrophoresis, using restriction endonuclease Bsp-digested λ dv 1 DNA as a marker.

DNA Reassociation Experiments. The renaturation of two samples of each plant DNA was monitored in a Gilford spectrophotometer 210, using *E. coli* DNA as a reassociation standard in each experiment. The application of 20, 120, and 400 mM NaPB allowed observation of DNA reannealing within a total Ecot-range of $10^{-5} <$ Ecot $\leq 1,500$, yielding adequate coincidence of the Ecot-curves in intervals of overlapping cot-values. Each experiment was reproduced one or more times, using plant DNA derived from various isolations. The amount of non-reassociated DNA at any cot-value was determined from absorption by a formula of Wetmur & Davidson (1968), which has been fitted to the output of the Gilford plotting-unit, 1,100. The absorbance data at 260 nm were corrected to allow for the volume expansion of water, according to the Gay-Lussac law. A collapse hyperchromicity of single stranded DNA amounting to 3% was applied for a further correction. This estimate is well in agreement with the value of Britten & al. (1974) for DNA of a GC-content of 40 Mol%, normally found for higher plant DNAs. Measuring the collapse hyperchromicities in distilled water, which does not permit DNA reannealing, produced comparable results. The data of the various DNA reannealing experiments were used to compute the fractions f and the reassociation constants k for a maximum of three components, renaturing in a bimolecular reaction of second order, by means of a "least square fit", which makes use of the formula:

$$D = \sum_{i=1}^{i=3} \frac{f_i}{1 + k_i f_i \,\mathrm{Ecot}}; \quad \sum_{i=1}^{i=3} f_i = 1, \text{ where } D \text{ is the}$$

amount of non-reassociated DNA at a given Ecot-value. The reassociation kinetic parameters listed in Tables 1 and 2 were calculated by the formulas given by BRITTEN & al. (1974), normalizing the redundancy r of the unique fractions to $r = 1$. The analytical complexity of the E. coli genome was assumed to be 4.23×10^6 Bp (CAIRNS 1963).

Abbrevations: Bp = base pairs; Ecot = equivalent cot; HAP = hydroxyapatite; NaPB = sodium phosphate buffer; NT = nucleotides.

DNA Interspersion Experiments. Radioactively labelled tracer DNA of various fragment lengths was renatured with a vast excess (1:5,000) of short driver DNA (average size: about 400 NT). These experiments were carried out in siliconized microcaps (50 μl). Palindromic reassociation was allowed to occur in 20 mM NaPB (Ecot-values $< 10^{-5}$). Renaturation of repetitive sequences was performed under conditions, which permit only repetitive DNA to react (up to an Ecot of 100, using 400 mM NaPB). The incubation temperature was chosen to be 25 °C below the T_m-value of native DNA, as demanded by WETMUR & al. (1968). After reannealing renatured and denatured DNA were separated by HAP chromatography according to a standard method (60 °C; elution of single-stranded DNA with 80 mM NaPB as necessary for both plant DNAs). The elution volume (16 × column volume) was shown to be sufficient for complete elution of single-stranded DNA. The fraction R of reassociated DNA, bound to HAP was calculated by:

$$R = \frac{\mathrm{cpm_{column}} - \mathrm{cpm_{summarized\ over\ single\text{-}stranded\ fractions}}}{\mathrm{cpm_{column}}}$$

where $\mathrm{cpm_{column}}$ is the total radioactivity loaded onto the column. The portion B of reannealed repetitive DNA bound to HAP was corrected for zero-time-binding DNA Z according to

$$B = \frac{R - Z}{1 - Z} \text{ (DAVIDSON & al. 1973)}.$$

Since reassociated plant DNA forms large networks — hyperpolymeres (FLAVELL & SMITH 1975) — it is necessary to cumpute the fractions of renatured DNA by the amounts of single-stranded DNA eluted from the column. DNA hyperpolymeres cannot be eluted from HAP under standard conditions. For this reason the recovery of reassociated DNA bound to HAP becomes more and more restricted with increasing DNA tracer lengths. The percentage of reassociated DNA bound to HAP was plotted against the average fragment size of the DNA tracers (see Figs. 4 and 5).

Results and Discussion

The genome of *M. incana* consists of four DNA fractions: a palindromic, a very fast repetitive, a slow repetitive, and a unique component (see Fig. 1 and Tab. 1). The very same fractions can be

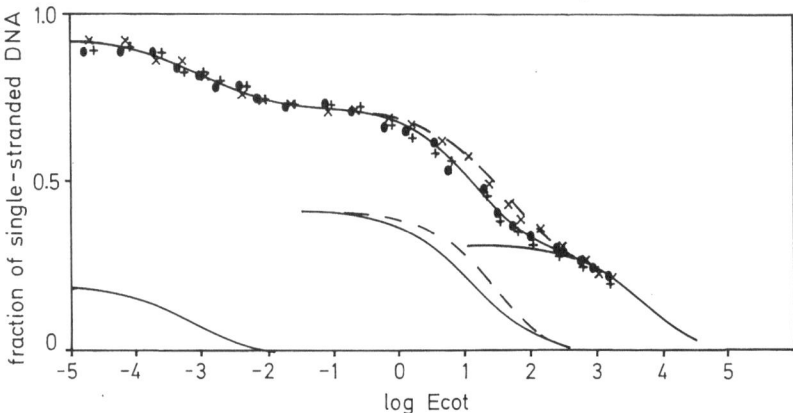

Fig. 1. Renaturation kinetics of the *Matthiola incana* genome (*Cruciferae*).
(———) total DNA; (- - - -) nuclear DNA. Average fragment length of both
DNAs: 450 Bp

Table 1. The reaccosiation kinetic components of the *Matthiola incana* genome,
strain 17 (*Cruciferae*). — 1 C-value of *M. incana* DNA: 1.5 pg or 1.37×10^9 Bp;
$cot_{1/2pure} = f \times cot_{1/2}$; f, fraction of component; DNA fragment length: 450 Bp;
the kinetic complexities are not corrected for sequence mismatch and
GC-content. $cot_{1/2}$(mole   NT × sec × 1^{-1});   reassociation   constant
$k = 1/cot_{1/2}(1 \times mole\,NT^{-1} \times sec^{-1})$

component	fraction	$cot_{1/2}$	$cot_{1/2}$ pure	k	k pure	kinetic complexity in Bp	number of copies per 1 C-genome
palindromes	0.070						
highly repetitive DNA	0.201	7.463×10^{-4}	1.50×10^{-4}	1,340	6,667	283	973,000
slow repeats total DNA	0.420	4.318	1.814	0.2316	0.551	1.637×10^6	352
nuclear DNA	0.415	8.573	3.558	0.117	0.281	2.955×10^6	192
single copy total DNA	0.309	1.116	345	8.958×10^{-4}	2.898×10^{-3}	4.233×10^8	1
nuclear DNA	0.314	1,248	392	8.012×10^{-4}	2.55×10^{-3}	4.749×10^8	1

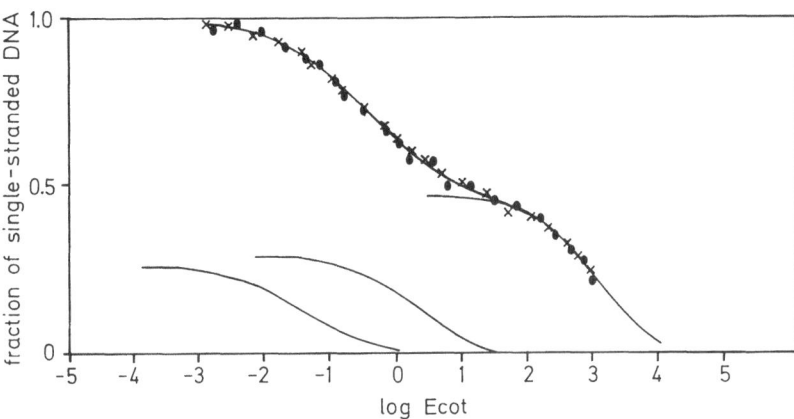

Fig. 2. Renaturation kinetics of the *Brassica pekinensis* genome (*Cruciferae*). Average fragment length of DNA: 450 Bp

Table 2. The reassociation kinetic components of the *Brassica pekinensis* genome cv. "Granat" (*Cruciferae*). — 1 C-value of *B. pekinensis* DNA: 0.505 pg or 4.61×10^8 Bp; $\text{cot}_{1/2 \text{ pure}} = f \times \text{cot}_{1/2}$; f, fraction of component; DNA fragment length: 450 Bp; the kinetic complexities are not corrected for sequence mismatch and GC-content. $\text{Cot}_{1/2}$ (mole NT $\times$ sec $\times$ 1^{-1}); reassociation constant $k = 1/\text{cot}_{1/2}$ ($1 \times$ mole $\text{NT}^{-1} \times \text{sec}^{-1}$)

component	fraction	$\text{cot}_{1/2}$	$\text{cot}_{1/2}$ pure	k	k pure	kinetic complexity in Bp	number of copies per 1 C-genome
palindromes	0.03—0.04						
fast repeats	0.249	1.151×10^{-2}	2.866×10^{-3}	86.9	349	5.07×10^3	22640
slow repeats	0.285	0.510	0.145	1.961	6.88	2.24×10^5	587
single copy	0.470	493	232	2.03×10^{-3}	4.31×10^{-3}	2.17×10^8	1

found in the DNA of *B. pekinensis* (see Figs. 2 and 3 and Tab. 2). The fast repeats of *M. incana* exhibit some special features. They consist of very highly repetitive DNA out of perhaps only one sequence class, which is reiterated nearly a million times per haploid genome. Such a high degree of repetition has not yet been found in the DNA of higher plants. This may be due to the fact that nearly all plants so far investigated were cultivated plants such as *Triticum aestivum* (FLAVELL & SMITH 1976), *Gossypium hirsutum* (WALBOT & DURE 1976), *Nicotiona tabacum* (ZIMMERMAN & GOLDBERG 1977), *Glycine max* (GOLDBERG 1978), and *Petroselinum sativum* (KIPER & HERZFELD 1978). All these plants seem to lack very fast repetitive sequences (a million or even more copies per genome), such as can be found in a greater number of mammals, e.g. the mouse genome (CORNEO & al. 1970). Another

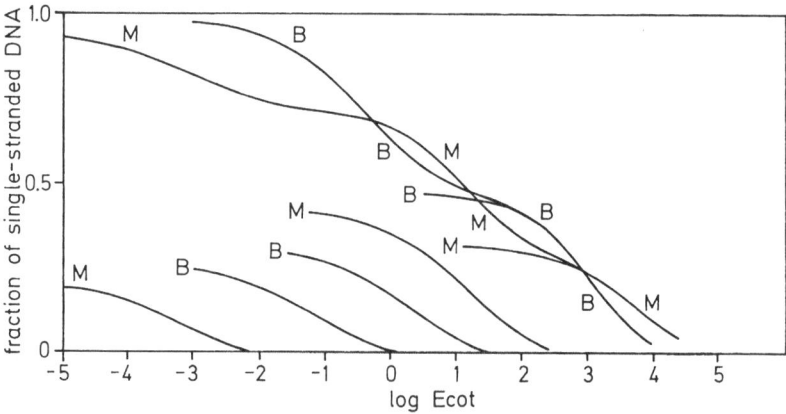

Fig. 3. Renaturation kinetics of the *Brassica pekinensis* and the *Matthiola incana* genome in comparison. M = *M. incana*; B = *B. pekinensis*. Average fragment length of both DNAs: 450 Bp

reason may be that highly repetitive sequences cannot be detected by HAP-chromatography in reassociation kinetic experiments, which does not allow a distinction between these sequences and inverted repeats. Nevertheless, the high percentage of this component in *M. incana* remains surprising (see Fig. 1 and Table 1). No significant reduction in the T_m-value can be observed for *M. incana* DNA after reannealing to an Ecot of 10^{-2}, indicating an almost perfect base pairing of the fast repeats. This phenomenon cannot be explained by the sequence mismatch of a GC-rich component, since HEMLEBEN & al. (1977) could not discover significant quantities of satellite or even cryptic satellite DNA by density centrifugation methods. The fast repeats of *M. incana* may be composed of simple sequence DNA organized within a satellite-like pattern. This sequence arrangement can be concluded from the reassociation of longer DNA fragments (1,900 Bp and greater 4,000 Bp in length), yielding a constant amount of this component (data not shown; optical monitoring of DNA renaturation). Nevertheless, the sequence organization of this DNA component should be studied in more detail by interspersion experiments.

The differences between nuclear and total DNA of *M. incana* could be caused by cytoplasmic DNA of mitochondria and chloroplasts, because of the different numbers of copies within the slow repeats (see Fig. 1 and Table 1), although the amount of extranuclear DNA (approximately 0.5 %) will hardly be significant.

The minicot curves of *M. incana* repetitive and unique DNA show adequate coincidence with the computed cot-curves, plotted in Fig. 1.

In contrast to the DNA of *M. incana*, which represents a more primitive wild crucifer species, the genome of the related, highly developed and cultivated species *Brassica pekinensis* lacks such a very fast repetitive component (see Fig. 2 and 3, and Table 2). For an explanation, possible trends in the phylogenesis of crucifers may be proposed:

1) Evolution favoured the diversification or the loss of the highly repetitive sequence.

2) The original intermediate-repetitive fraction split into two classes of reiterated sequences (fast and slow repeats, not so strongly differing in redundancy).

In particular, this divergence into fast repeats, consisting of a smaller number of related sequence families, might allow a highly integrated regulation of genes, coding for functionally related proteins. On the other hand, the great number of slightly different sequence classes within the slow repeats could take care of a more specific regulation of single genes or smaller groups of genes. The anagenesis of crucifers may perhaps result from such a better cooperativity of gene regulation, increasing the adaptibility of plants suitable for cultivation. The much more "*Brassica*-like" DNA composition of all cultivated plants investigated so far is consistent with this assumption. A greater number of crucifers should be compared with respect to the fast repeats, in order to support or to disprove this speculation on genome evolution in this family.

Further data on the sequence arrangement within the *M. incana* genome can be obtained from Figs. 4 and 5, which represent the results of the interspersion experiments. The straight line in Fig. 4 cuts the axis at an extrapolated value of 7.3 %. This is well in agreement with the palindromic fraction computed by reassociation kinetics (see Table 1). Since this line cuts the axis at a positive value, the inverted repeats in *M. incana* DNA must be arranged in regularly spaced clusters (HAMER & THOMAS 1974). This seems to be a general feature of plants, as

3*

it was also found in wheat (Flavell 1976) and parsley
(Kiper & Herzfeld 1978), and animals (Wilson & Thomas 1974).
According to the palindrome theory of Hamer & Thomas (1974) an
average spacer length between palindromic clusters of 3,300 ± 400 NT
(simple standard deviation) can be calculated. From the intercept of
the line in Fig. 4 an average size of palindromic clusters of

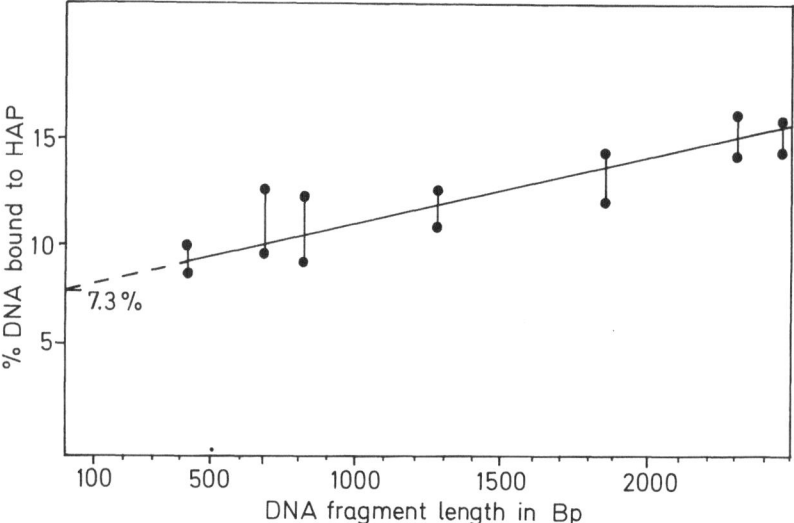

Fig. 4. Tracer length dependence of HAP binding of *Matthiola incana* zero-time
binding DNA

approximately 300 Bp can be predicted. S_1-nuclease digestion of
palindromic reassociated DNA (DNA fragment size 5,000 NT; stringent
buffer conditions as described by Davidson & Britten 1973) yields S_1-
resistant DNA fragments shorter than 100 Bp in length (determinated
by agarose gel electrophoresis), indicating S_1-digestion at single-
stranded inflection-sites within palindromic clusters.

From the extrapolated line in Fig. 5 it can be concluded that about
57 % of the repetitive sequences are interspersed. Thus about 12 % of
the total repetitive fraction (69 %; see Table 1) may be organized in
clustered repeats; probably these are the highly repetitive sequences
forming the presumed satellite-like pattern of simple sequence DNA,
which does not affect the intercept of the extrapolated line. Therefore,
the amount of clustered repeats may comprise 12 or even more percent

(up to the portion of the highly repetitive sequences). This further gives rise to the necessity of HAP interspersion studies of the highly repetitive component in *M. incana* DNA. The average size of redundant heteroduplices can be estimated tp 300—400 Bp by means of hyperchromicity data of reannealed repetitive DNA.

S_1-nuclease resistant fragments of repetitive DNA (DNA fragments

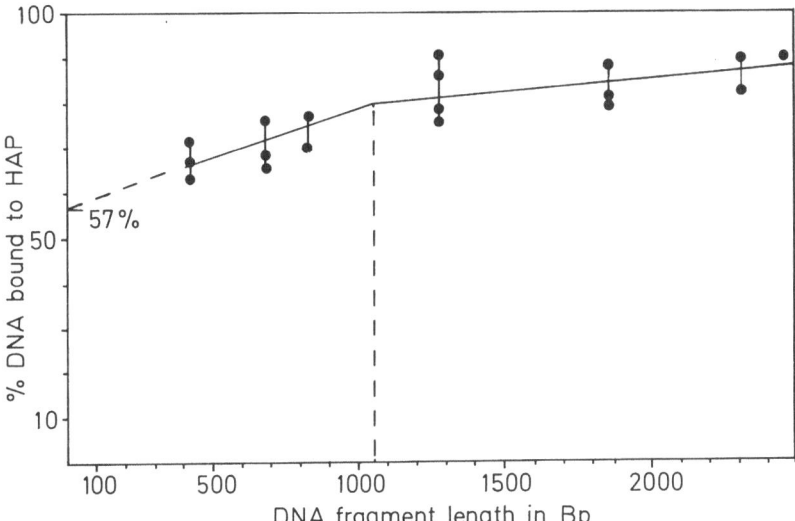

Fig. 5. Tracer length dependence of HAP binding of *Matthiola incana* repetitive DNA

longer than 5,000 NT were renatured to cot 100; S_1-digestion was performed under stringent buffer conditions) yielded a peak at about 110 Bp in agarose slab gels. Only a small percentage of the S_1-resistant repetitive duplices seem to be longer than 400 Bp, up to a maximum of about 1,150 Bp. Beyond 400 Bp there is a continuos increase in digested fragment length, reaching the peak mentioned (data not shown). This could result from the sequence mismatching of repetitive sequences of about 300 to 400 Bp in length. Further experiments varying the S_1-digestion conditions must, therefore, be carried out, in order to obtain more precise results. The repetitive sequences are separated by an unique spacer of an average size of about 1,050 NT (see Fig. 5). Hence, a *Xenopus*-like interspersion pattern seems to be predominant in the *M. incana* genome. A sequence arrangement of this kind is suggested to

be a general one in higher plant DNA (Flavell & Smith 1975). Up to
now this has been confirmed in all plants investigated, e. g. in a number
of cereal species (Rimpau & al. 1978), soybean (Goldberg 1977),
tobacco (Zimmerman & Goldberg 1977), and parsley (Kiper &
Herzfeld 1978). Approximately 20% of the repetitive sequences
are separated by longer unique segments. Thus, the sequence organization of the *M. incana* genome can be compiled as follows:

fraction *f*

1) clustered repeats
   (perhaps a satellite-like pattern)   12—20%
2) a *Xenopus*-like interspersion pattern   60—68%

$$-\left[\frac{300\text{—}400\,\text{Bp} \quad 1{,}000\text{—}1{,}100\,\text{Bp}}{\text{repetitive} \qquad \text{single copy}}\right]_n*$$

3) Other patterns (unique clusters;
   non-*Xenopus*-like patterns)   20%

It can be concluded that about $n = 600{,}000$ *Xenopus*-like organized
segments must be present in the *M. incana* genome. Similar numbers
can be computed for almost all plants investigated. Given that uniques
are protein-coding units, one should perhaps understand these
Xenopus-like arranged segments as potential genes. But only about
10,000 or slightly more genes are found to be active in a series of plants
(Gollmer & Hemleben 1979, Kiper & al. 1979, Goldberg & al. 1978).
On the other hand, a genetic capacity of more than 100,000 genes (still
considerably smaller than the estimated number of "gene-like"
segments) would drive an organism into genetic instability, because of a
well established mutational rate of 10^{-5} per gene per generation. The
phenomenon of "gene-excess" may be explained by some simple
biological facts:

1) Selection and further evolutionary factors act upon phenotypes,
that is to say upon products of active genes, and not upon the genes
themselves.

2) The selection of phenotypes occurs during early stages of
embryogenesis, whereas postnatal selection may be neglected. This is a
well established fact in human genetics and animal breeding.

3) According to Haeckel's principle, prenatal selection must act
upon evolutionary intermediates, that is upon ancient proteins.

---

* $n = \dfrac{f \times \text{genome size}}{\text{segment length}} \approx \dfrac{0.64 \times 1.37 \times 10^9}{1{,}500} \approx 600{,}000$

HAECKEL's principle will therefore be a prerequisite of prenatal selection.

In molecular terms, these facts should mean that HAECKEL's fundamental principle must be structurally encoded in the eukaryotic DNA sequence organization. Each component of the eukaryotic genome, and not merely the repetitive sequences, would therefore contribute to the phenomenon of DNA excess (C-value paradox), which is therefore a "gene-excess". For testing this hypothesis, protein patterns from early stages of development and from evolutionary intermediates should be compared.

The determination of the DNA content of *M. incana* and *B. pekinensis* was carried out by Dr. H. GAUDE, Institute of Biology III, University of Tübingen.

References

BRITTEN, R. J., KOHNE, D. E., 1968: Repeated sequences in DNA. — Science **161**, 529—540.

— GRAHAM, D. E., NEUFELD, B. R., 1974: Analysis of repeating DNA sequences by reassociation. In GROSSMAN, L., MOLDAVE, K., Eds.: Methods in Enzymology. Vol. **29 E**, pp. 363—418. — New York: Academic Press.

CAIRNS, J., 1963: The chromosome of *E. coli.* — Cold Spring Harb. Symp. Quant. Biol. **28**, 43—46.

CORNEO, G., GINELLI, E., POLLI, E., 1970: Different satellite DNAs of guinea pig and ox. — Biochemistry **9**, 1565—1570.

DAVIDSON, E. H., HOUGH, B. R., AMENSON, C. S., BRITTEN, R. J., 1973: General interspersion of repetitive with non-repetitive elements in the DNA of *Xenopus.* — J. Molec. Biol. **77**, 1—23.

— BRITTEN, R. J., 1973: Organization, transcription, and regulation in the animal genome. — Quart. Rev. Biol. **48**, 555—613.

FLAVELL, R. B., SMITH, D. B., 1975: Genome organization in higher plants. Stadler Symp. Vol. **7**, University of Missouri, Columbia.

— — 1976: Nucleotide sequence organization in the wheat genome. — Heredity **37**, 231—252.

GOLDBERG, R. B., 1978: DNA sequence organization in the soybean plant. — Biochem. Genetics **16**, 45—68.

— HOSCHEK, G., KAMALAY, J. C., TIMBERLAKE, W. E., 1978: Sequence complexity of nuclear and polysomal RNA in leaves of the tobacco plant. — Cell **14**, 123—131.

GOLLMER, I., HEMLEBEN, V., 1979: Transcriptional activity in seedlings of *Matthiola incana* determined by DNA-RNA hybridization. — Pl. Syst. Evol., Suppl. **2**, 151—161.

HAMER, D. H., THOMAS, C. A. JR., 1974: Palindrome theory. — J. Molec. Biol. **84**, 139—144.

HEMLEBEN, V., GRIERSON, D., DERTMANN, H., 1977: The use of equilibrium centrifugation in actinomycin-caesium chloride for the purification of ribosomal DNA. — Pl. Sci. Lett. **9**, 129—135.

KIPER, M., HERZFELD, F., 1978: DNA sequence organization in the genome of *Petroselinum sativum (Umbelliferae).* — Chromosoma **65**, 335—351.

Kiper, M., Bartels, D., Köchel, H., 1979: Gene number estimates in plant tissues and cells. — Pl. Syst. Evol., Suppl. **2**, 129—149.

Nagl, W., 1976: Zellkern und Zellzyklen. — Stuttgart: Eugen Ulmer.

Rimpau, J., Smith, D., Flavell, R. B., 1978: Sequence organization analysis of the wheat and rye genomes by interspecies DNA/DNA hybridisation. — J. Molec. Biol. **123**, 327—359.

Walbot, V., Dure, L. S. III., 1976: Developmental biochemistry of cotton seed embryogenesis. VII. Characterization of the cotton genome. — J. Molec. Biol. **101**, 503—536.

Wetmur, J. G., Davidson, N., 1968: Kinetics of renaturation of DNA. — J. Molec. Biol. **31**, 349—370.

Wilson, D. A., Thomas, C. A. jr.: Palindromes in chromosomes. — J. Molec. Biol. **84**, 115—144.

Zimmerman, J. L., Goldberg, R. B., 1977: DNA sequence organization in the genome of *Nicotiana tabacum*. — Chromosoma **59**, 227—252.

Address of the authors: Dipl.-Biol. Wolfgang Wenzel and Prof. Dr. Vera Hemleben, Institut für Biologie II, Lehrstuhl für Genetik der Universität Tübingen, Auf der Morgenstelle 28, D-7400 Tübingen, Federal Republic of Germany.

Pl. Syst. Evol., Suppl. 2, 41—66 (1979)

Premedical Biology Program, University of
Heidelberg, Federal Republic of Germany

Genome Size and Phenotypic Evolution in
Microseris (*Asteraceae, Cichorieae*)

By

Konrad Bachmann, Heidelberg, Kenton L. Chambers, Corvallis,
and H. James Price, College Station

Key Words: Angiosperms, *Asteraceae = Compositae, Cichorieae, Microseris.*
— Evolutionary genetics, genome size, developmental canalization, polymorphism, heterocarpy.

Abstract: Reduction of the nuclear DNA amount in a constant number of
n = 9 chromosomes characterizes the evolutionary genetics of the genus
Microseris. Correlated with this reduction are physiological, ecological and
morphological changes. The morphological changes are of two kinds: (1)
nucleotypic, i.e. changes in the dimensions correlated with DNA amount
changes; (2) qualitative, specifically concerning an increase in the degree of
developmental canalization with a decrease in genome size. The latter changes
are amenable to formal genetic analysis, and two polygenic systems are
described in some detail. As far as preliminary estimates indicate, the
complexity of these polygenic systems is independent of genome size. They are
probably influenced by genome size via nucleotypic limitations on gene
expression.

Each species of animal or plant has a characteristic nuclear DNA
amount which varies within such narrow limits among different
specimens of any one species that it seems to be determined and
maintained by natural selection. In striking contrast to this intraspecific
constancy closely related species may have entirely different genome
sizes (summaries: Sparrow & al. 1972, Bennett & Smith 1976). There is
usually no obvious reason why any species should have the genome size
it has, and at least the difference in DNA amount between closely similar
species appears to be so much superfluous DNA (Ohno 1972).

The combination of intraspecific constancy and interspecific
variability suggests that an evolutionary approach based on the

comparison of closely related species with differing nuclear DNA
amounts may be the natural way to shed light on this apparent waste of
nuclear DNA. A number of authors have made such comparisons (see
PRICE 1976, also NAGL & EHRENDORFER 1974; GREILHUBER, this volume).
Their observations show that the nuclear DNA amount is anything but
unimportant. Rather, it seems to play a central role in the determination
of the specific phenotype of a species as well as in the evolutionary
processes that change or maintain the species-specific phenotype. This
quantitative role of the nuclear DNA amount as opposed to the
qualitative role of the information content of the genome, has been
characterized most strongly by BENNETT (1971, 1972), who introduced
the term "nucleotype" for the quantitative "mechanical" properties of
the nuclear DNA and placed these effects side by side with the ones of the
genotype and the environment. BENNETT (1972) lists the following
phenotypic characters of herbaceous plants which are correlated with
the nuclear DNA content: chromosome volume, nuclear volume, cell
size, nucleolar and nuclear dry mass, seed dry mass, minimum cell cycle
time, meiotic duration, pollen maturation time and minimum generation time. The various characters of this list can be summed up under
two headings. They concern either the relative timing of developmental
or growth events or the relative sizes of the structures formed in these
events. The nucleotype therefore is concerned with the scaling in time
and space of events which are qualitatively determined by the genotype.
This description of the role of the nuclear DNA amount is undoubtedly
simplified but it serves as a clear statement on which to base further
discussion.

The relatively simple mode of growth of higher plants lets these
scaling effects influence the total phenotype more directly than this is
the case in animals. In animals, a direct correlation between nuclear
DNA amount and cell size can be demonstrated easily, but this
correlation is practically never carried over to organ or body size. On the
other hand, comparative studies of animal species with different nuclear
DNA amounts have revealed another correlation which complicates the
discussion considerably. This correlation concerns the degree of
morphological specialization of species in relation to their nuclear DNA
content. As first illustrated by HINEGARDNER (1968) in fish, those species
with particularly low DNA amounts are morphologically most deviant
from the typical stream-lined fish phenotype and represent the end
points of specialized lines of evolution. Similar correlations have been
found in several vertebrate and invertebrate groups, and they can also
be found in plants (BACHMANN & al. 1972, PRICE 1976). In all these cases,
evolution leading from a more generalized ancestral type to morphologically more specialized forms is accompanied by a reduction in the

nuclear DNA amount. Why this should be the case is by no means evident.

A mere quantitative "mechanical" DNA effect is unlikely to be accompanied by specific qualitative changes. Anyway, the common correlation between evolutionary reduction in DNA amount and morphological specialization allows no predictions about the kinds of specializations which are evolved. The reduction in nuclear DNA amounts does not lead to one predictable specialized phenotype. Rather it seems to favor the achievement and the maintenance of a specialized phenotype, whatever the specialization may be. It appears impossible to explain this observation without involving that part of the DNA which carries genetical information. There may be a loss of structural genes involved in morphological specialization, there may be a loss of regulatory DNA sequences, in any case, the DNA lost during evolution must include or at least influence structural genes in order to have the kind of qualitative effect which accompanies its loss. For this reason it becomes important to consider genotypic effects together with nucleotypic effects in DNA evolution. A formal genetic analysis applied to characters which evolve in coordination with the nuclear DNA amount should show how the DNA loss makes itself felt genetically.

In the following we shall show that such an analysis is feasable, and we shall demonstrate some of the genetical peculiarities of the morphological characters involved. We have as yet not a single firm answer to the question we are analyzing. Still, already the preliminary part of this project has shown that the genetics of morphological specialization correlated with evolutionary DNA loss touches on central problems of evolutionary genetics which have never been properly investigated.

Material and Methods

The subtribe *Microseridinae* of tribe *Cichorieae* (*Compositae*) contains a group of closely related species which constitute an ideal system for the investigation of genome evolution. The genus *Microseris* contains about 16 species (Table 1; CHAMBERS 1955, 1957) of which five are tetraploid ($2n = 4x = 36$). Except for one species in Chile and one in Australia and New Zealand, the genus is limited to Western North America. Together with related species in the genera *Phalacroseris*, *Nothocalais*, and *Agoseris* there are about 30 species of which the evolutionary relationships are fairly well established on the basis of morphology, distribution and the results of experimental hybridization experiments. This number of species is large enough to allow generalizations to be drawn from it and sufficiently small to be studied in great detail. In all of these species the haploid chromosome number is $x = 9$ (Fig. 1). Among the diploid species of *Microseris* genome sizes vary over a threefold range. Including *Phalacroseris* and *Agoseris* provides a variation of genome sizes of seven to one (PRICE & BACHMANN 1975). The diploid species of *Microseris* can be arranged in a series according to their genome size. This arrangement correlates well with the

Table 1. Taxonomic survey of the genus *Microseris*: figures in parentheses are picograms of DNA per haploid genome

1) Subgenus *Scorzonella* (NUTT.) SCH.-BIP.
   diploid perennials, North America

   M. laciniata (HOOK.) SCH.-BIP.   (3.4)
   M. howelii GRAY
   M. paludosa (GREENE) J. T. HOWELL   (3.1)
   M. nutans (HOOK.) SCH.-BIP.   (3.5)
   M. sylvatica (BENTH.) SCH.-BIP.   (3.4)

2) Subgenus *Apargidium* (TORR. & GRAY) K. CHAMBERS
   diploid perennial, North America

   M. borealis (BONG.) SCH.-BIP.   (4.0)

3) Subgenus *Moniermus* (HOOK. f.) K. CHAMBERS
   tetraploid perennial, Australia, New Zealand

   M. scapigera (SOL. ex A. CUNN.) SCH.-BIP.   (5.8)

4) Subgenus *Microseris*
   a) diploid annuals, North America

   M. lindleyi (DC.) A. GRAY   (2.0)
   M. douglasii (DC.) SCH.-BIP.   (1.4)
   M. bigelovii (GRAY) SCH.-BIP.   (1.5)
   M. elegans GREENE ex GRAY   (1.4)
   b) diploid annual, Chile
   M. pygmaea D. DON   (1.4)
   c) tetraploid annuals, North America

   M. decipiens K. CHAMBERS   (*lind. + big.*)
   M. heterocarpa (NUTT.) K. CHAMBERS   (*lind. + dougl.*)
   M. campestris GREENE   (*eleg. + dougl.*)
   M. acuminata GREENE

postulated evolutionary history of the genus. Starting out with fairly tall perennial outcrossing ancestral species of mesic woodlands for which *M. laciniata* serves as a model, evolutionary specialization coupled with genome size reduction finally leads to small, highly self-fertile annuals of open grasslands. For these, *M. elegans* is an extreme example.

Natural populations of *Microseris* species have been sampled in California and Oregon. From these collections, representative specimens of all major biotypes have been grown in the greenhouse. From a few of these, inbred laboratory populations have been established. Field surveys of selected populations have been made in spring and summer of 1978. These surveys consisted of smapling usually one mature head of seed per plant of a representative sample of plants. Accessions are identified by their CHAMBERS(CH) or their PRICE & BACHMANN (PB) numbers. Voucher specimens are preserved in the herbaria of Oregon State University.

The following characters were scored in individual mature heads of fruit: number of phyllaries, separate number for inner and outer phyllaries where separable, and number of achenes. Of each achene the number of pappus parts was determined as well as the color, the presence or absence of a spotting pattern, the presence or absence of "hairiness", and the fertility as far as it could be determined by inspection (well-developed *vs.* collapsed achenes).

Thus the material consists of population surveys (variability at one site), biotype surveys based on a few selected individual specimens from different

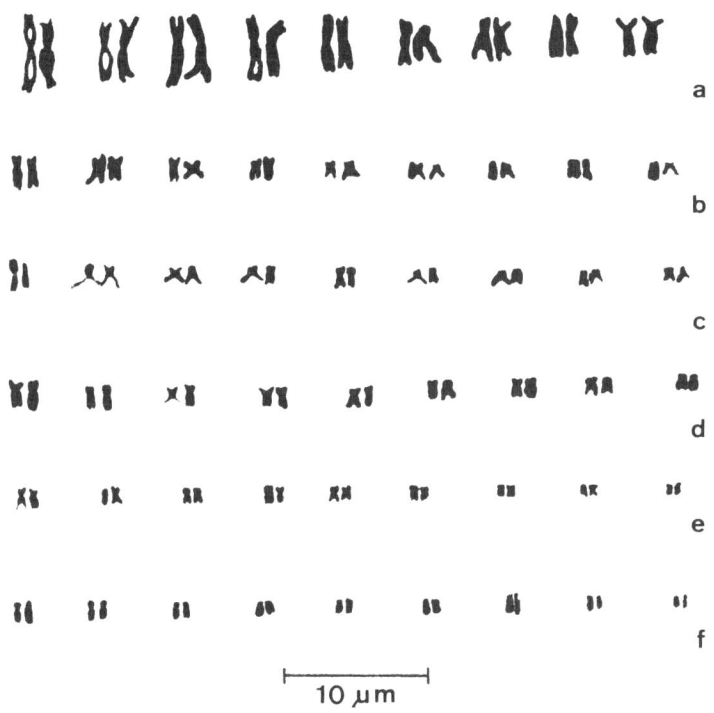

Fig. 1. Representative karyotypes of six microseridine species. *a Phalacroseris bolanderi, b Microseris laciniata, c M. sylvatica, d M. nutans, e M. lindleyi, f. M. douglasii*

populations grown side by side in the greenhouse (range of variability within the species), and the analysis of inbred lines derived either from individual plants collected in nature or from greenhouse hybridizations.

Nature of the Phenotypic Change

Most of the quantitative characters which we have measured in different species of *Microseris* are subject to scaling: Their major (though by no means only) fate during the evolution of specialized annual species from generalized perennials has been a reduction in size or number. At the same time, there are qualitative changes which lead to a more specialized phenotype. Both size reduction and specialization can be detected in most characters, and there isn't a single one in which a qualitative phenotypic change doesn't interfere with a straightforward

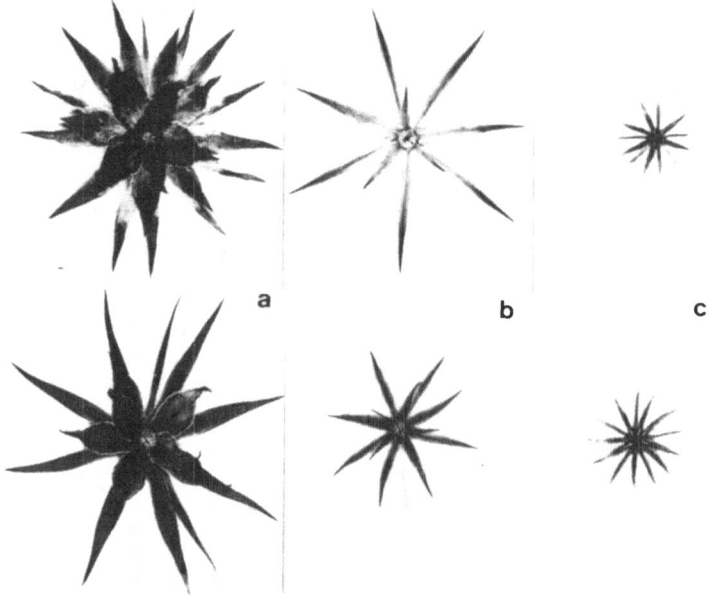

Fig. 2. Involucres, pressed flat, viewed from below, all photographed to the same scale. *a M. laciniata*, primitive phenotype above, more canalized type with three sets of phyllaries $(7 + 8 + 2)$ below. *b* Extremes of natural size variation in *M. lindleyi* and *c M. elegans*

quantitative analysis. The structure of the involucre surrounding the flowering head illustrates this well (Fig. 2). The advanced annuals have clearly smaller phyllaries. This is the scaling effect. At the same time, there is a qualitative trend to differentiate the phyllaries into two distinct sets, a set of longer inner ones and a set of scale-like outer ones. This is an example for a process of qualitative morphological specialization. The generalized involucre consists of a spirally arranged series of phyllaries which gradually change in shape, the specialized involucre consists of two circles of phyllaries with an abrupt morphological switch from one type to the other. The basic feature of this evolutionary process is the replacement of continuous gradual morphological change during individual development by discontinuous sudden switches. This qualitative tendency is as obviously correlated with the reduction in genome size as the change in dimensions. The data which we have collected up to now on the evolution of the involucre do not yet warrant a detailed discussion. It is clear, however, that there is genetic variation for involucre characters within species. Variation of phyllary length is particularly striking in *M. laciniata* and *M. lindleyi*. Some of this variation undoubtedly is genetically determined, even though we are not yet able to estimate the heritability of phyllary length. The qualitative structure of the involucre also varies within species, and the offspring of one of our laboratory plants of *M. laciniata* (60105) segregates into plants with more generalized and plants with more specialized involucres.

Quantitative Characters Subject to Scaling

An analysis of intraspecific versus interspecific variation in any one of the characters changing quantitatively during the evolution of *Microseris* should indicate how much of the genetic component of this variability is genome-size independent ("genotypic") and how much of it is genome-size dependent ("nucleotypic"). In principle such an analysis is straightforward. In practice, so many sources of random and non-random variation affect the expression of any continuously variable character that we are far from able to calculate reliable genetic variances. More simply, it is virtually impossible because of all the different sources of variability to determine a species-specific mean-value for characters such as achene weight or number of flowers per head in such a way that the values are comparable among species. As easy as it is to see clear differences of size or number among the various species, any measurement of these can at best be semi-quantitative. One method of obtaining rather reliable species-specific characteristics is to use derived measures based on the correlated variation among various direct measures. In that case variability need not be minimized but is made use of. An example for this is represented by Fig. 3. In this figure, average achene weights for individual laboratory grown specimens of *Microseris* are plotted against average achene lengths. The samples are very small and very biased for special environmental conditions. The species averages of either measure therefore are very unreliable. However, the correlation between achene length and achene weight shows strikingly that the evolution of achene shape from *M. laciniata* through *M. elegans* is mainly a process of scaling. If we calculate for each species a shape characteristic based on the data, the values are very similar (Table 2).

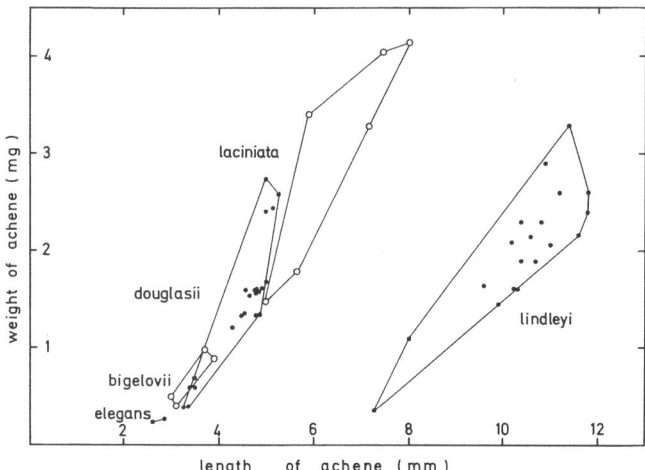

Fig. 3. Weight versus length of the achene without pappus of five species of
Microseris; laboratory populations; each point represents the average value for
a single specimen

Table 2. Achene weight in mg divided by the third power of achene length in
mm. This constant depends primarily on the shape of the achene

Species	Strain	a	$\pm$ s	n
M. laciniata	60	0.0111	0.0029	6
M. douglasii	26	0.0152	0.0028	15
M. douglasii	37	0.0131	0.0023	5
M. douglasii	all	0.0149	0.0027	22
M. bigelovii	25	0.0160	0.0023	4
M. elegans	24	0.0121	0.0012	2
all above		0.0142	0.0030	34
M. lindleyi	22,30	0.0020	0.0010	20

Essentially in this group of species the achenes differ by their dimensions
more than by anything else. The data for *M. lindleyi* show how sensitive
this method is. Even the few data presented here leave no doubt that the
achene of *M. lindleyi* is qualitatively different from the typical
Microseris achene. This difference can be expressed quantitatively as a
decrease by a factor of 7 in the shape constant (Table 2).

   This provides us with a method to distinguish quantitative scaling
effects from qualitative changes and to measure them separately. Using

this approach, intraspecific variability can be compared to the distance between species. Intraspecific variability for achene shape and size exists most strikingly in *M. douglasii*. This has been discussed in detail by CHAMBERS (1955). The observations have not yet been supplemented by quantitative measures comparable to those represented in Fig. 3. Still, both the intraspecific genetic variability and the methods to measure and analyze it are available.

Pappus Part Number

General Observations. One process of morphological specialization in the evolution of *Microseris* suggested a very simple genetical correlation with genome size, when we began investigating it. This concerns the number of pappus parts on the achenes (corresponding to the number of sepals per flower). This number is essentially constant in the specialized annuals. Usually it is five. In the one Chilean species of *Microseris*, *M. pygmaea*, it is ten. Contrasting with this constancy in the annuals is a variability of the pappus part numbers among the achenes of a single head of *M. laciniata*. *M. laciniata* from near Dos Rios, California (CH 2960 = PB 60) shows any number of pappus parts between five and ten on achenes of virtually every single head.

Our obvious first assumption was that the reduction of genetic variability which resulted in the canalization of pappus part numbers might be a direct consequence of the loss of genetic material from a polygenic system. The very first genetical data on our laboratory populations of the annuals (BACHMANN & CHAMBERS 1979), and a garden population of *M. laciniata* PB 60 (BACHMANN & PRICE 1979), however, excluded any such simple correlation between genome size and phenotypic canalization. The genetics of pappus part number turned out to be very complex, and genetic variability far exceeding that found in *M. laciniata* PB 60 can be obtained in an annual, when *M. pygmaea* with its ten-part pappus is crossed with a five-part species like *M. bigelovii* (Fig. 5; BACHMANN & CHAMBERS 1978). The new results concern primarily a survey of species and biotypes which was performed with the aim of outlining the total variability contained in the polygenic system determining pappus part number within the genus. That there is considerably more than the variability documented for our laboratory populations is evident already from the pappus part numbers mentioned by CHAMBERS (1955, 1957).

Annuals. Most of the phenotypic variability in the pappus system of annual species is exhibited by *Microseris douglasii*. This species is the phenotypically most variable annual in every respect.

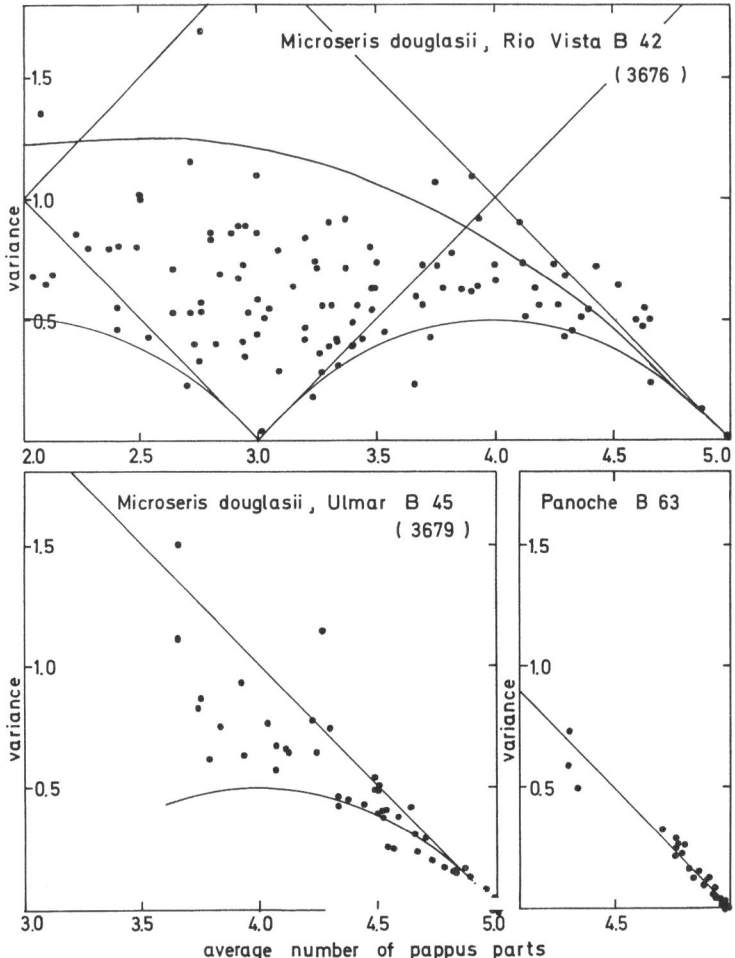

Fig. 4. Natural variation of pappus part numbers in three California populations of *M. douglasii*. Each point represents the distribution from a single head of one plant as expressed by average and distribution variance. Theoretical lines delimit predictions based on Poisson and binomial distributions. B 42: near Rio Vista, Solano County; B 45: near Livermore Laboratory, Alameda County; B 63: Panoche Hills, Benito County

The variability of *M. douglasii* differs among populations. There may be a regular clinal pattern, but that remains to be established. Certainly limited local stands (topodemes) of *M. douglasii* differ considerably in the amount of phenotypic variation they contain. This variation concerns all three parameters needed to characterize the distribution of pappus part numbers (Fig. 4): the basic number and the direction and magnitude of the average deviation. By far most of the plants of *M. douglasii* have a basic number of five pappus parts. A nearly pure three-part pappus pattern has been found in a specimen of the Rio Vista population (CH 3676, PB B42). Three as basic number is present together with five in quite a few specimens of this population. It is evident in all those distribution curves canalized between 5 and 3 instead of 5 and 10 or 5 and zero. The statistical analysis of population B42 also indicates basic numbers of 1 and/or zero, but no plant bearing exclusively single-part pappi or having only apappose achenes has yet been found. For all other California annuals five is the only basic number of pappus parts.

The direction of the deviation from the basic number of 5 is most frequently to lower numbers. Ideally, such a "5—" type produces achenes with five, four, three etc. pappus parts with frequencies predicted by a Poisson distribution for numbers missing from five (Table 3). When the basic number of three sets a lower limit to such a distribution, an effect is observed, which we have discovered already in the artificial hybrids between *M. pygmaea* and *M. bigelovii* (BACHMANN & CHAMBERS 1978), where "10—" distributions are stopped at the lower limit of 5. In the douglasii population B42 many plants are found in which only achenes with 3, 4, and 5 pappus parts are found. The resulting distribution curves rarely allow to decide if these are "3 +" types with 5 as upper limit or "5—" types with three as lower limit. A thrilling possibility is suggested by the data, and is amenable to experimental analysis. This concerns the bracketing of the limits of canalization for the pappus part number. It may be that a "5—, 3 +" type results in a distribution of pappus parts caught between three and five, while a corresponding "5—, 3—" type allows the "5—" distribution to run through to lower numbers. We have some pieces of evidence supporting such a model. Foremost among them is the segregation of our laboratory population PB 26 of *M. douglasii*. This population seems to segregate for "5 +" and "5—" types, indicating that the direction of the deviation from the basic number may be determined by different alleles of a single Mendelian factor (see below, Table 4). The nature of such a gene would be of considerable interest. For this reason we are trying hard to map the two morphological markers which segregate together with the "5—" allele. The families of PB 26 tested up to now have turned out to be homozygous. Still, we are well on our way to get a formal genetic analysis of limit and direction of canalization.

The average amount of the deviation from the basic number presents

4*

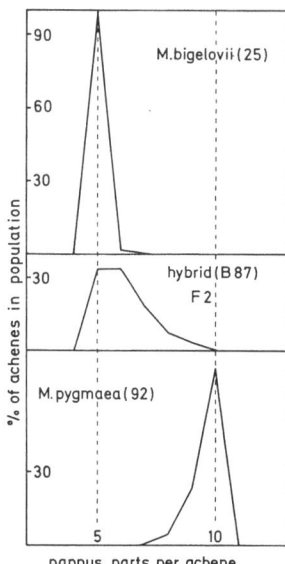

Fig. 5. Frequency distribution of achenes with different pappus part numbers in *M. bigelovii*, *M. pygmaea*, and the F 2 of a *pygmaea* × *bigelovii* hybrid. The strong canalization of pappus part numbers of the parents breaks down in the hybrid, which still shows broad canalization between 5 and 10 and a dominance effect for low numbers

Table 3. *M. elegans*, specimen CH 3722-3, Jolon, Monterey County, California. Frequency distribution of pappus part numbers. Basic number 5, average deviation — 1.0

Pappus parts	0	1	2	3	4	5
Number found	0	0	28	145	293	302
Number expected	3	12	47	141	283	283

a much more complicated problem. There may be very many genotypes and consequently very many genes involved in its determination. The one indication that the number of genes may be limited and exactly countable is the frequent occurrence of round numbers. Thus, an average deviation of unity from a basic number is not only found in the artificial hybrids between *M. pygmaea* and *M. bigelovii* (BACHMANN & CHAMBERS 1978) but also in specimens of *M. douglasii* collected in nature. While the other annuals usually have much smaller deviations from the basic number of five pappus parts, we have raised one specimen of *M. elegans* from an achene collected near Jolon, California (specimen CH 3722-3) which represents a rather clear "5—1" type (Table 3). It may be noted that as far as we have determined, *M.*

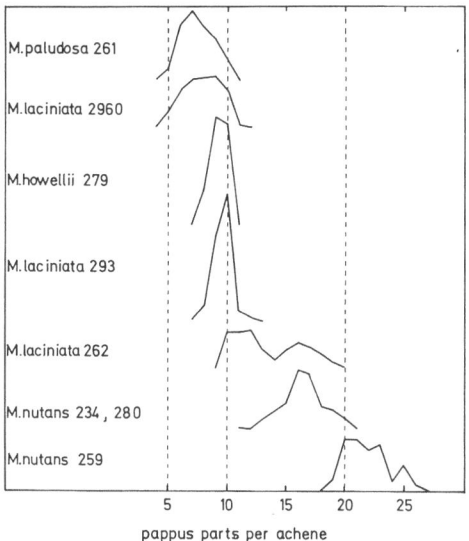

pappus parts per achene

Fig. 6. Representative distributions of pappus part numbers in eight topodemes of perennial *Microseris* species. Accessions identified by Chambers locality numbers. *M. laciniata* 2960 (= PB 60) is subspecies *laciniata*, 293 subsp. *leptosepala*, 262 a subspecies from Jackson Co., Oregon; *M. nutans* 280 and 259 are the same subspecies from localities in Josephine Co., Oregon and Del Norte Co., California, resp.

elegans is predominantly of the type "5—", while its close relative, *M. bigelovii*, is predominantly "5 +". *M. pygmaea*, the third species in this genetically very similar group, is always "10—".

Perennials. Genetic experiments with the perennial species of *Microseris* require much longer time periods than those with annual species. The following analysis therefore is based mainly on the analysis of populations collected in nature, supported by some very preliminary data on first-generation species hybrids.

Basic numbers of pappus parts in the perennials are 5, 10, 20, and 40. This is illustrated by the population distributions shown in Fig. 6, where the numbers 5, 10, and 20 clearly are distribution limits. 40 seems to be a limit in *M. borealis*. These numbers immediately suggest a polygenic system with multiplicative effects. The few first-generation species hybrids which have been analyzed for their pappus part numbers support this view and add another aspect (Fig. 7). Hybrid mean values either are very near the geometric mean of parental averages, as would be expected with a multiplicative system, or the hybrid is very strictly canalized on a basic number between the parental means.

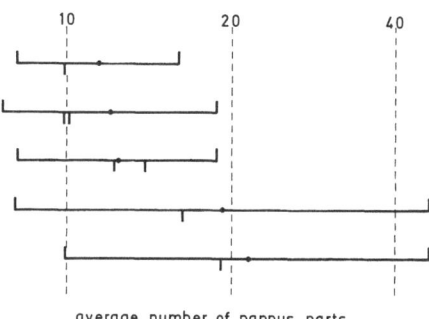

average number of pappus parts

Fig. 7. Average pappus part numbers in seven F 1 hybrids from five interspecific crosses of perennial *Microseris* species: parental averages upward marks, offspring average downward marks, points indicate geometric means of parental averages; logarithmic scale

We want to point out that high pappus part numbers obviously are a derived character for the perennials, while low numbers are the derived character for the annuals. The ancestral character state therefore most likely is a distribution between five and ten as seen in *M. laciniata* PB 60 or a firmly canalized phenotype at 10 or 5. A more detailed knowledge of the genetics of this character will probabely eliminate any ambiguity and show exactly which way the evolutionary changes in this genetic system have gone and in which way the different species are related to each other. There remains no doubt that the genetic system determining pappus part number is homologous throughout the genus, that it is complex, that it has evolved in parallel with and as part of the morphological evolution of the genus and that an elucidation of its details are just a matter of time.

The "Hairy Achenes" System

In scoring the phenotype for any of the characters mentioned one soon realizes that quantitative variability within a single flowering head is linked with the geometry of the head. The major feature of the

Genome Size and Evolution in *Microseris*   55

geometry of the composite inflorescence is the spiral arrangement of
phyllotactic sites. All structures appearing sequentially at homologous
sites along this "genetic spiral" can be ordered in a developmental
sequence from outer earlier ones to inner later ones. In simple cases this
spiral can be followed from the basal rosette of leaves up the stalk
supporting the head where typical leaves are more and more replaced by
bracts and finally by the phyllaries of the involucre. From there the
developmental spiral continues through the attachment sites of the

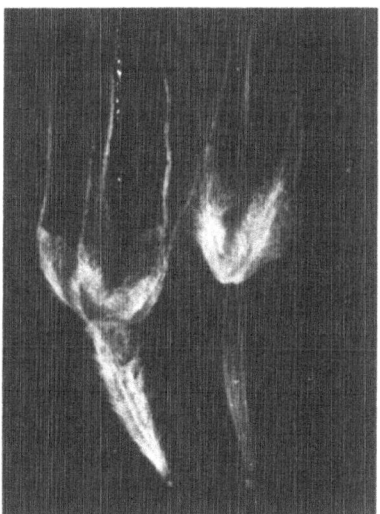

Fig. 8. *Microseris douglasii*, specimen PB 26115, marginal "hairy", and central
achenes; achene length 4 mm

achenes. The evolutionary trend described for the involucre above
concerns the serial homology along this spiral. A gradual morphological
change from one homologous structure to the next is replaced by a
discrete stepwise change at one specific point along the spiral. The
transition from phyllaries to achenes in principle constitutes another
such switch, and we have preliminary indications that by developmental
accident (and, possibly, by mutation) the site (or time) at which the
switch takes place gets shifted. As a result, there are in occasional heads
too many achenes relative to the number of phyllaries or *vice versa*.

Among the different achenes of a single head, gradual or stepwise
differentiations along the developmental spiral are the rule. If striking
morphological differences among different achene types result, the plant

is said to be heterocarpous. Heterocarpy in *Microseris* can concern differences in overall shape, in color, and in surface structure of the achenes. Since surface structure can be scored in all achenes of every head including aborted sterile fruit, we have examined heterocarpy based on surface structure differences as a probe for developmental switching genes.

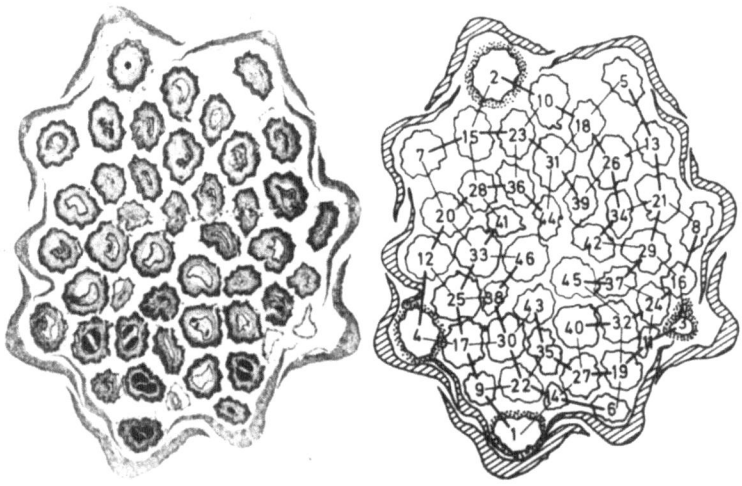

Fig. 9. Cross section through mature capitulum of an annual *Microseris*; there are eight (heavy lines) and 13 (thin lines) major parastichies defining the sequence of the achenes; achenes 1 through 4 are hairies, achenes 3, 11, and 14 are early lethals. Specimen B 35330, *pygmaea* × *elegans* hybrid F 3

In most species, with the conspicuous exception of *M. lindleyi*, outer achenes may be "hairy" (Fig. 8). This phenotype results from the loosening of cell wall adhesions so that individual cells protrude causing a fuzzy to densely furry appearance of the achene surface. This surface structure can be seen even in collapsed aborted fruit, and it can be seen in cross-sections of maturing heads. This allows to map the positions of hairy achenes on the head and to relate them exactly to the developmental spiral (Fig. 9; cf. POMPLITZ 1956).

We have scored the percentage of hairy achenes among all achenes and used this as a quantitative character representing a developmental switch along the phyllotactic spiral. Undoubtedly it is not directly that percentage which is genetically controlled and a more detailed analysis

Table 4. *M. douglasii* PB 26, segregation for pappus part number, achene spotting, violet stripe on phyllaries, and percent of hairy achenes

Specimen	Number of pappus parts						Average	achenes	Phyllaries	% Hairies	N
	3	4	5	6	7	8					
26104			22				5.000	spotted		5	22
26105		5	233				4.979	spotted		17	238
26101			103				5.000	brown		18	103
26127		2	47				4.837	spotted	green	22	49
26118	1	16	162				4.899	spotted	green	23	181
26108	7	30	232				4.836	spotted	green	25	261
26111	5	24	196				4.849	spotted	green	28	208
26112	5	29	291				4.880	spotted	green	30	325
26117		1	86	1			5.000	brown	striped	33	88
26120			330	3			5.036	brown	striped	34	328
26102			62				5.000	brown		35	62
26115			162	5		1	5.048	brown	striped	35	168
26121			244	1			5.004	brown	striped	36	245
26103			161	5	2	1	5.071	brown		36	233
26114		10	163	1			4.948	brown	striped	39	173

of the development of the capitulum will turn up a better way to score the same effect with less variability in the data from a single plant. Still, the percentage of hairy achenes has turned out to be another quantitative polygenically determined character for which homologous genetic systems are present in most *Microseris* species.

Again, *M. douglasii* turns out to be the most polymorphic annual with plants frequently having no hairy achenes at all, rarely nothing but hairy achenes and otherwise a great many different possible percentages between zero and one hundred. This character is virtually independent of environmental influences except indirectly by its slight dependence on the total number of achenes per head. This leaves enough of a variability to cause difficulties in assigning all the phenotypes found to specific genotypes. Again, though, a complete analysis of this polygenic system is only a matter of time, and we have already several hints as to individual genotypes and their non-genetic variability. One of these comes from segregation in *M. douglasii* PB 26 (Table 4). The same morphological markers which segregate together with the "5 + /5—" gene also are linked with genes for the percentage of hairy achenes. In this case, offspring from homozygous plants, even though it does not contribute to the analysis of linkage, at least serves to delimit individual genotypes within this polygenic system.

M. campestris, the tetraploid derivative of *M. douglasii* and *M. elegans*, helps in the identification of genotypes for the "hairy"-system. We have found only three very clearly separated phenotypes in this species up to now. Population PB B46 collected near Livermore, Alameda County, California, consists of two types of plants, those having no hairy achenes and those having between 5 and 7 percent of them. Population B55 collected in the Panoche Hills in Fresno County also contains only two types of plants, those with no hairy achenes and those with 10 to 16 percent of them. A survey of eleven individual plants from five other localities has turned up nine times zero percent and once each four and twelve percent. Even a conservative estimate has to reckon with at least ten phenotypes for this character in the annuals, possibly many more.

M. laciniata and its subspecies among the perennials may have between zero and one hundred percent hairy achenes, usually though with a less pronounced degree of hairiness. Population CH 4523 collected south of Grant's Pass, Josephine County, Oregon, comprises plants with between zero and 42 percent hairy achenes. Of forty plants, 27 had between zero and two percent hairy achenes, 12 between nine and 17, and one had 42. This distribution suggests a HARDY-WEINBERG equilibrium for two alleles of a single gene, with 82 % "zero-Alleles" and 18 % "42-Alleles".

Our standard population of *M. laciniata*, PB 60 has been scored for hairy achenes only recently. The offspring of specimen 60105 segregates for no hairies and 23% hairies. All indications are that the number of different phenotypes for this system found in the annuals and in the perennials may be similar. There certainly is no indication that the larger genome of *M. laciniata* contains more genes for this character than the smaller genome of *M. douglasii*.

Discussion

This project is the first systematic attempt to study the evolutionary genetics of genome size reduction in terms of formal genetic analysis and thereby to connect two approaches of evolutionary genetics which up to now have had little contact. These are the investigation of genome evolution by comparative DNA analyses and the population genetic approach based on Mendelian genetics. There is an urgent need for a connection between the two approaches, since either alone will ultimately be unable to yield a complete understanding of evolutionary genetics. Analyses of DNA amounts and the evolution of DNA fractions have shown up a quantitative DNA effect on the phenotype (the "nucleotypic" effects) which seems to play a very prominent role in determining species differences. This effect is still being ignored in population genetics, where different species are essentially treated as allelic permutations of a basic set of homologous cistrons. Paradoxically, genetic material missing from the genome of a species may determine more of its distinguishing characteristics than those sequences held in common with its ancestors. On the other hand, mere shifts in dimensions and timing during ontogenetic development can not explain the evolution of species either. The relative influence nucleotype and genotype on the phenotype and their role in speciation and the evolution of higher categories has to be determined, and the result is likely to have a profound influence on evolutionary genetics.

Obviously the genus *Microseris* is as good a group of organisms for such an investigation as one can possibly expect. The evolutionary trend from a *laciniata*-like ancestor to the present-day annuals has involved the loss of more than half of the genome, and this loss involves all DNA fractions (BACHMANN & PRICE 1977; Table 5). At the same time the phenotype has changed in a way that immediately suggests a correlation between DNA loss and morphological evolution. Lastly, as shown here, genetical polymorphisms can be demonstrated for individual characters that take part in this morphological trend.

The data presented here have contributed no new information to the understanding of straight nucleotypic effects. Dimensional scaling can be demonstrated, measured, and quantitatively distinguished from qualitative effects. The precise determination of species-specific measures presents however so many practical and theoretical problems that we can not yet refine the correlation between DNA amount and any quantitative character among the various species. It appears from looking at the plants that the differences in scale among them are greater than the differences in total DNA amount. This could either be explained by a non-linear relationship or by a linear correlation of the nucleotypic characters with some fraction of the DNA which is reduced more than proportionally when the genome is pared down in evolution.

Table 5. Summary of genome structure data. Total DNA with calibration by BENNETT (1976), renaturation data BACHMANN & PRICE (1977), ribosomal RNA genes HEMLEBEN & al. (1978)

| | Total | Genome size (picograms) | | | Cistrons of | |
| | | Renaturation | | Fractions | r DNA | 5 S DNA |
	(haploid)	fast	inter-mediate	slow		
M. bigelovii	1.5	0.20	0.61	0.73	1970	
M. douglasii	1.4	0.29	0.34	0.76	3000	9.000
M. elegans	1.4				1100	
M. laciniata	3.4	0.68	1.16	1.56	3200	13.700
M. lindleyi	2.0	0.39	0.70	0.90	1560	10.000

The possibility that selected DNA fractions are responsible for all or some of the nucleotypic effects has first been suggested by the measurements of PEDERSEN (1971), who found a closer correlation between erythrocyte size and the number of ribosomal cistrons than between erythrocyte size and the total nuclear DNA amount in a comparison of five fish species. We have some data on the number of ribosomal cistrons in *Microseris species* (HEMLEBEN & al. 1978; Table 5) which do not support such a correlation for the nucleotypic characters of the genus. The problem, however, is still wide open. Puzzling in this respect is the enormous variability of *M. douglasii*. If the correlation between nuclear DNA and nucleotypically determined morphological characters is in any way close, then one should expect a corresponding intraspecific variability of the total DNA amount in this species. Selecting extreme biotypes on the basis of their morphology and then checking these for variations in the nuclear DNA amounts seems to be the most practical approach to the problem of intraspecific DNA amount variation.

While many groups of plants or animals could serve equally well for the investigation of straight quantitative nucleotypic effects, the genus *Microseris* is uniquely suited for the investigation of the relationship between "morphological specialization" and the nuclear DNA amount. HINEGARDNER (1968) was the first to document such a correlation by his classical series of outline drawings of fish species with various DNA amounts. Meanwhile similar correlations have been documented for several animal and plant groups. *Microseris* allows to translate the subjective impression of "morphological specialization" into quantitatively measureable terms. Thus it opens the way for an experimental analysis of this puzzling effect. The results of this analysis should help to find out if we are justified to generalize from the evolution of a genus of plants to that of bats and sea horses. For the moment, *Microseris* is the only system in which such an analysis is feasible.

The common feature to all these cases of "morphological specialization" is the correlation between a generalized phenotype and an intermediate DNA amount (both in relationship to the ranges represented within a group of related species), while small genome sizes are typical for species with derived, specialized, even outright bizarre morphologies (HINEGARDNER 1968). At the same time, there usually are more species with small genomes, and the process of species formation itself seems to be favored by a small genome size (BACHMANN & al. 1972). This suggestion is supported by the observation that very large genome sizes are typical for another kind of specialization which is found in species-poor groups with slow rates of evolution. A model for such a species might be the monotypic genus *Phalacroseris*, which is obviously a close relative of *Microseris* but combines a much higher nuclear DNA amount (Fig. 1) with morphological and ecological peculiarities.

Specifically in *Microseris*, "morphological specialization" means replacing a continuous variability in ontogenetic processes with stepwise changes. The result is a higher degree of canalization of characters in the low-DNA species. We have seen this in the structure of the involucre, we have seen it in the determination of pappus parts, and we can see it in the differentiation of achene types along the developmental spiral if we consider more characters than the one we have scored here.

We consider this the most significant result of this investigation thus far: analyzing the evolutionary trend leading to the annual species of *Microseris* we have found it to be a matter of control of ontogenetic development and of phenotypic canalization.

The second important result concerns the demonstration of

ubiquitous genetical polymorphisms for the characters involved. Even
with the limited genetical data which we have collected up to now, it is
striking that we find genetical variability wherever we look for it. Some
of this genetic variability is not translated into phenotypic variability in
nature. We had to hybridize *M. pygmaea* with *M. bigelovii* to reveal how
much genetical variability for the strictly canalized number of pappus
parts is hidden in the genome of *M. bigelovii* (BACHMANN & CHAMBERS
1978). Much of the genetical variability, even for canalized characters, is
seen as phenotypic variability in natural populations. The demonstration of this variability provides a new focus for the discussion of
the relative roles of nucleotypic and genotypic influences in species
differences.

The involucre of *M. laciniata* may serve as an illustration. In this
species we find the most generalized involucres within the genus, in which
there is a large number of phyllaries which show a gradual transition
from nearly leaf-like outer ones to elongated inner ones (Fig. 2). At the
same time there is very much variation for involucre structure within
and between populations of this one species, and we know that both
number and shape of the phyllaries are genetically determined. Stable
differences between plants raised for years side by side in the same
environment (BACHMANN & PRICE 1978), population differences, and the
first data on segregation among the offspring of one plant allow a
preliminary estimate about the genetical basis of this variation. From
this we may safely predict that a few generations of selection could
produce an involucre in *M. laciniata* as canalized and smaller on average
than that of *M. lindleyi* without any detectable change in nuclear DNA
amount. The kind of qualitative morphological change which accompanies the reduction of DNA amount in evolution should be obtainable
without this reduction. What, then, explains the relationship between
the reduction in DNA amount and the increasing morphological
specialization?

We are left with three possibilities which have to be explored by
further experiments:

1) There may be no developmental correlation between DNA
reduction and morphological specialization. In this case, the observed
correlation would be based on a parallel selection in two genetical
systems which do not influence each other in development. The fact that
similar correlations are found everywhere in the animal and plant
kingdom would be the result of an unwarranted homologization of
processes which have no common genetic basis.

2) The correlation may be direct. This is one of our early ideas based
on the transition from broad canalization of pappus part number in *M.
laciniata* to narrow canalization in the annuals. If this effect is due to the

loss of genes from a polygenic system, then ophenotypic effect and DNA loss would concern the same DNA sequences.

This idea at present appears very unlikely. Our estimates of the number of genes involved in the polygenic system which we have identified are very vague. Still, in both systems, the pappus part system and the hairy achenes system, there is not the least indication that the annuals contain fewer genes than *M. laciniata*. If anything, the results in both cases point the other way.

There is an intriguing modification of the direct-relation hypothesis. A polygenic system influences the phenotype not only by the number of its components but also by their individual effects. The amount added to a quantitative character by each allele of an additive polygenic series is as important to the expression of the phenotype as the number of alleles which are present. Now this strength of the individual allele may be dosis-dependent on the amount of DNA in the individual gene of the polygenic series. Thus each identifiable "gene" in a polygenic series could be a series of repeated cistrons under common control, with the number of cistrons in the unit determining its effect on the phenotype. This is a highly speculative idea. The advantage of *Microseris* is that it allows to test such ideas. We have already identified a multiplicative series determining the range of canalization of pappus parts in the perennials. Its unit strength is a doubling effect based on the number 5. We can analyze this doublin in the perennial hybrids. In addition, *M. scapigera*, the tetraploid species of Australia and New Zealand, seems to be polymorphic for this multiplicative system (SNEDDON 1978). Very soon, too, we shall be able to measure the strength of individual alleles in the hairy achenes system, and we shall be able to analyze a large body of data as soon as we learn how to count genes in the canalized pappus part system.

3) The third possible explanation for the connection between DNA loss and morphological change postulates an indirect effect. This is the most likely connection on the basis of the available data. According to this hypothesis, DNA loss, as far as it has a phenotypic effect is completely independent of the informational role of the DNA. On the other hand, there is no loss of DNA in the genetic systems determining the morphological characters. These systems are homologous troughout and are polymorphic to various degrees and for various alleles with little or no relationship to genome size. The interaction between the two effects is in development: The phenotypic expression of a specific homologous genotype changes from one species to another, because nucleotypic characters provide a different developmental environment for the action of the same genes.

For the pappus part numbers of *M. laciniata* BACHMANN & PRICE

(1978) have identified two opposing influences oriented along the radius of the developing capitulum. These may correspond to two opposing diffusion gradients of substances determining high and low pappus part numbers respectively. It can easily be imagined that such a gradient system would interact in a completely different way if set up across a much smaller capitulum. The nucleotypically determined conditions of meristem size, cell size, cell division rate etc. could easily modulate the action of genes.

These three hypotheses show that we are still far from an understanding of the genetics of genome size. The data on *Microseris* presented here show, however, that this genus provides an ideal system for the analysis of these problems. Already the initial results suggest a new paradigm for evolutionary genetics which agrees with everything we know but adds some essential aspects which have been neglected up to now.

Summary

1) The evolution of the diploid species of *Microseris* leads from morphologically more generalized perennials of mesic woodlands to specialized self-fertile annuals of dry grasslands.

2) Correlated with this trend is a decrease in nuclear DNA amount which involves all fractions assayed, while the chromosome number remains constant at n = 9.

3) The phenotypic changes which are correlated with DNA loss in evolution are studied with the aim of finding the common genetic mechanism.

4) Straight "nucleotypic" changes concern a genome-size correlated reduction in the dimensions of organs. This can be demonstrated qualitatively. A strong dependence on environmental factors and the interaction between quantitative and qualitative changes impede quantitative analysis.

5) Qualitative changes in the phenotype correlated with DNA reduction predominantly concern the degree of developmental canalization of characters. Gradual changes in homologous structures along the phyllotactic spiral are replaced by abrupt switches from one expression to another. This is illustrated in the evolution of the involucre.

6) Pappus part number determination is used as a system to investigate progressive canalization. The genetics of this character are complex and three components of canalization are under independent control: the limits, the direction of deviation from the limits, and the magnitude of the deviations. Inter- and intra-specific variability for all

components can be demonstrated, and the genetic complexity of the system can be roughly estimated.

7) The switch from "hairy" to "smooth" achenes along the phyllotactic spiral of achene arrangement serves as a model for the investigation of developmental switches. The timing of this event is scored as the percentage of hairies among all achenes. This character is polygenically determined. Preliminary estimates of non-genetic variability allow a rough estimate of the genetic complexity of the system.

8) There is no indication of a direct involvement of these polygenic systems in the DNA loss. The most likely connection between these polygenically determined canalizations in development and genome size is an influence of genome-size dependent dimensional scaling on the expression of the polygenes in development.

This research is supported by grants Ba 536/1-4 from the *Deutsche Forschungsgemeinschaft* and DEB-7809940 from the National Science Foundation. Additional support for field work was provided by Oregon State University through the kindness of dean R. KRAUSS. We acknowledge with thanks the help of Professor P. SCHNEIDER (Fig. 8), Mr. J. SACK (the photo of Fig. 9), and Ms. G. SACK, who contributed Fig. 1 and a determination of the nuclear DNA amount of *Microseris paludosa*. Mr. H. BACK, Mr. A. KONRAD, and Ms. E. OELDORF performed most of the arduous task of scoring the field collections.

References

BACHMANN, K.. HARRINGTON, B. A., CRAIG, J. P., 1972: Genome size in birds. — Chromosoma (Berl.) **37**, 405—416.

— CHAMBERS, K. L., 1978: Pappus part number in annual species of *Microseris (Compositae, Cichoriaceae)*. — Pl. Syst. Evol. **129**, 119—134.

— GOIN, O. B., GOIN, C. J., 1972: Nuclear DNA amounts in vertebrates. — Brookhaven Symp. Biol. **23**, 419—450.

— PRICE, H. J., 1977: Repetitive DNA in *Cichorieae (Compositae)*. — Chromosoma (Berl.) **61**, 267—275.

— — 1979: Variability of the inflorescence of *Microseris laciniata (Compositae: Cichorieae)*. — Pl. Syst. Evol. **131**, 17—34.

BENNETT, M. D., 1971: The duration of meiosis. — Proc. Roy. Soc. (Lond.) **B 178**, 277—299.

— 1972: Nuclear DNA content and minimum generation time in herbaceous plants. — Proc. Roy. Soc. (Lond.) **B 181**, 109—135.

— SMITH, J. B., 1976: Nuclear DNA amounts in angiosperms. — Phil. Trans. Roy. Soc. (Lond.) **B 274**, 227—274.

CHAMBERS, K. L., 1955: A biosystematic study of the annual species of *Microseris*. — Contrib. Dudley Herbarium **4**, 207—213.

— 1957: Taxonomic notes on some *Compositae* of the western United States. — Contrib. Dudley Herbarium **5**, 57—68.

HEMLEBEN, V., BACHMANN, K., PRICE, H. J., 1978: Ribosomal gene numbers in *Microseridinae (Compositae: Cichorieae)*. — Experientia **34**, 1452—1453.

HINEGARDNER, R., 1968: Evolution of cellular DNA content in teleost fishes. — Amer. Natur. **102**, 517—523.

NAGL, W., EHRENDORFER, F., 1974: DNA content, heterochromatin, mitotic index, and growth in perennial and annual *Anthemidea (Asteraceae)*. — Pl. Syst. Evol. **123**, 35—54.

OHNO, S., 1972: So much "junk" DNA in our genome. — Brookhaven Symp. Biol. **23**, 366—370.

PEDERSEN, R. A., 1971: DNA content, ribosomal gene multiplicity, and cell size in fish. — J. Exp. Zool. **177**, 65—78.

POMPLITZ, R., 1956: Die Heteromorphie der Früchte von *Calendula arvensis* unter besonderer Berücksichtigung der Stellungs- und Zahlenverhältnisse. — Beitr. Biol. Pflanzen **32**, 331—369.

PRICE, H. J., 1976: Evolution of DNA content in higher plants. — Bot. Rev. **42**, 27—52.

— BACHMANN, K., 1975: DNA content and evolution in the *Microseridinae*. — Amer. J. Bot. **62**, 262—267.

SACK, G., 1978: Cytologische Untersuchungen an verschiedenen *Cichorioideae (Compositae)*. Undergraduate thesis, Heidelberg.

SNEDDON, B. V., 1978: A biosystematic study of *Microseris* subgenus *Monermos*, *(Compositae: Cichorieae)*. — Dissertation, Victoria University, Wellington, New Zealand.

SPARROW, A. H., PRICE, H. J., UNDERBRINK, A. G., 1972: A survey of DNA content per cell and per chromosome of prokaryotic and eucaryotic organisms: some evolutionary considerations. — Brookhaven Symp. Biol. **23**, 451—494.

Address of the authors: Prof. Dr. KONRAD BACHMANN, Premedical Biology Program, University of Heidelberg, Im Neuenheimer Feld 504, D-6900 Heidelberg, Federal Republic of Germany.

Pl. Syst. Evol., Suppl. 2, 67—71 (1979)

Department of Biology, The University
of Kaiserslautern, Federal Republic of Germany

Repetitive DNA in Primitive Angiosperms

By

Maria E. Schaan and Walter Nagl*, Kaiserslautern

Key Words: Angiosperms, *Cinnamomum*, *Decaisnea*, *Liriodendron*, *Magnolia*. — DNA base composition, DNA repetition, genome evolution.

Abstract: The proportion of repetitive DNA was determined in four primitive angiosperms by analysis of the renaturation kinetics of DNA extracted from leaves. The genome of *Magnolia soulangiana* consists of 39.5% repetitive sequences, that of *Liriodendron tulipifera* 47.5%, that of *Cinnamomum camphora* 37.3% and that of *Decaisnea fargesii* 48.5%. The cot curves were similar in all species studied. The results confirm the hypothesis that the proportion of repetitive DNA increased during macro-evolution, and that primitive, or stable, taxa show less variation in the genome composition than more advanced, rapidly evolving taxa.

Repetitive DNA sequences are characteristic for eukaryotic genome organization (BRITTEN & KOHNE 1968). The proportion of repetitive DNA is apparently higher in more advanced taxa than in "primitive" taxa (reviewed by NAGL 1976): DNA of fungi displays, on average, 13% repetitive DNA, that of dicots 50-80%, and DNA of monocots 70-95% (for some details see FLAVELL & al. 1974, 1977). In some genera it has been found that the variation in repetitive DNA is the main reason for the observed variation in nuclear DNA contents between species (2 C values), and it was suggested that this variation in the amount of repetitive DNA played some role in speciation and splitting phenomena during evolution (NAGL 1978).

In this study we analyzed the proportion of repetitive DNA in four primitive angiosperms in order to answer the following questions:

1) Is there evidence for increase of the amount of repetitive DNA during macro-evolution? If this hypothesis is right, the slowly evolving

---

* To whom all correspondence should be addressed.

5*

"primitive" angiosperms should show considerable lower amounts than the rapidly evolving "higher" taxa.

2) Is there evidence for less variation in the proportion of repetitive DNA in the slowly diverging taxa than in the rapidly diverging modern taxa?

3) Is there evidence for a dependence of the 2 C value from the amount of repetitive DNA also among less closely related species?

Material and Methods

The following plants were used for the experiments: *Magnolia × soulangiana* Soul.-Bod. and *Liriodendron tulipifera* L. (both *Magnoliaceae, Magnoliales*), *Cinnamomum camphora* (L.) Sieb. (*Lauraceae, Magnoliales*), and *Decaisnea fargesii* Franch. (*Lardizabalaceae, Ranunculales*). The plants were grown in the Botanical Garden and green-houses of the university of Kaiserslautern; vouchers are preserved in the herbarium Klu. DNA was isolated from leaves according to the method of Wells & Ingle (1970), denatured in a thermocuvette equipped with a thermoprogrammer, and the hyperchromicity was recorded by a Gilford spectrophotometer. Renaturation of fragments of about 400–600 nucleotides in length was done at 60 °C in 0.12 M phosphate buffer. All experiments were repeated 3–6 times.

Results

Table 1 and Fig. 1 summarize the results. The melting points of the DNA's show only little variation. The base composition was calculated from the Tm's according to the formula given by Marmur & Doty (1962) and estimated to be between 35 and 39%. These are rather low values for angiosperms. *Decaisnea*, which is less related to the other species, shows a clear deviation from the values obtained among the *Magnoliales*.

The renaturation kinetics of the DNA's were transferred into cot curves which were evaluated according to the criteria given by Britten & al. (1974). Using *E. coli* DNA as a standard it was calculated that unique sequences reassociated at a cot1/2 of 16–67.

The proportion of repetitive DNA, which was estimated from the kinetic data for the four species, is clearly lower than in higher angiosperms, ranging between 37 and 49 %. The variation between the genera studied is smaller than the variation found between species of higher angiospermal genera (Flavell & al. 1974, Nagl 1976). With respect to individual cot fractions, the DNA of the four species is composed of 20 % highly repetitive DNA (cot 10^{-1}) with a redundancy between 1.26×10^3 and 8.58×10^4. The amount of intermediately repetitive DNA exhibits the greatest variation between the four species studied. The question, whether the proportion of repetitive DNA does

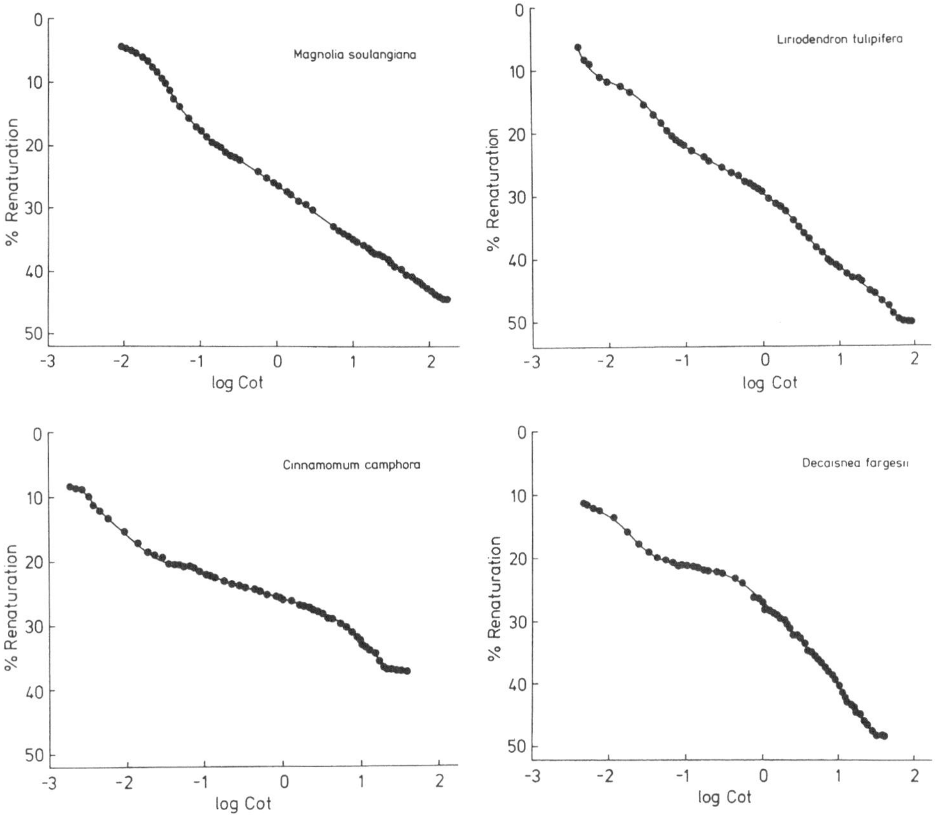

Fig. 1. Reassociation kinetics (cot curves) of the DNA from four primitive angiosperms (means of 3–6 experiments)

show a positive correlation with the nuclear DNA content (2 C value) or not, cannot be answered unequivocally. The data given in Table 1 indicate a positive correlation except in the case *Magnolia*. As *Magnolia*, however, must be considered an ancient polyploid genus (EHRENDORFER & al. 1968, NAGL & al. 1977), and because only a small number of species was studied, a positive correlation between DNA content and proportion of repetitive DNA may also exist in primitive angiosperms.

Conclusions

The questions asked at the beginning of this report can be answered in a preliminary mode only, because not more than four species were

Table 1. Characterization of the nuclear DNA from four pimitive angiosperms
(means of 3–6 experiments each)

Species	Hyper-chromi-city (%)	Tm[a] (°C)	GC (%)	Repet. DNA (%)[b]	Genome size (1 C) (pg)
Magnolia soulangiana	31.5	85.0	38.7	39.5	6.0[c]
Liriodendron tulipifera	38.7	84.6	37.7	47.5	0.8
Cinnamomum camphora	40.0	84.5	37.4	37.3	0.6
Decaisnea fargesii	35.5	83.6	35.2	48.5	2.5

[a] Tm = melting temperature, i. e. temperature at which half of the hyperchromicity is reached
[b] Proportion of total genome
[c] A generative-polyploid species

studied so far. In these representative primitive angiosperms, however, evidence was found for

1) a low proportion of repetitive DNA,

2) little variation of repetitive DNA between species, and

3) a positive correlation between 2 C value and proportion of repetitive DNA.

Thus, the results are consistent with the suggestion, that variable amounts and a variable composition of repetitive DNA are attributes of progressive taxa, because changes in the non-coding repetitive DNA may be the molecular basis of the control of morphogenesis, and thus of splitting and speciation (Nagl 1977, 1978, 1979).—Details of the study will be given in a forthcoming article.

Technical assistance by Miss B. Holzhauser is gratefully acknowledged. We thank the *Deutsche Forschungsgemeinschaft* for support of this work (grant NA-107/3), and Professor H. Huber, Kaiserslautern, for helpful discussion.

References

Britten, R. J., Kohne, E., 1968: Repeated sequences in DNA. — Science **161**, 529—540.
— Graham, D. E., Neufeld, B. R., 1974: Analysis of repeating DNA sequences by reassociation. — Meth. Enzymol. **29 E**, 363—406.
Ehrendorfer, F., Krendl, F., Habeler, E., Sauer, W., 1968: Chromosome numbers and evolution in primitive angiosperms. — Taxon **17**, 337—353.

FLAVELL, R. B., BENNETT, M. D., SMITH. J. B., SMITH, D. B., 1974: Genome size and the proportion of repeated nucleotide sequence DNA in plants. — Biochem. Genet. **12**, 257—269.

— RIMPAU, J., SMITH, D. B., 1977: Repeated sequence DNA relationships in four cereal genomes. — Chromosoma **63**, 205—222.

MARMUR, J., DOTY, P., 1962: Determination of the base composition of DNA from its thermal denaturation temperature. — J. Molec. Biol. **5**, 109—118.

NAGL, W., 1976: Zellkern und Zellzyklen. Stuttgart: Ulmer.

— 1977: The DNA optimization model for speciation and cytodifferentiation. — Chrom. Today **6**, 151—152.

— 1978: Endopolyploidy and polyteny in differentiation and evolution. Amsterdam: North-Holland.

— 1979: Search for the molecular basis of diversification in phylogenesis and ontogenesis. — Pl. Syst. Evol., Suppl. **2**, 3—25.

— HABERMANN, T., FUSENIG, H.-P., 1977: Nuclear DNA contents in four primitive angiosperms. — Fl. Syst. Evol. **127**, 103—105.

WELLS, R., INGLE, J., 1970: The constancy of the buoyant density of chloroplast and mitochondrial DNA's in a range of higher plants. — Pl. Physiol. **46**, 178—179.

Address of the authors: MARIA ELISABETH SCHAAN, Prof. Dr. WALTER NAGL, Department of Biology, The University, P.O. Box 3049, D-6750 Kaiserslautern, Federal Republic of Germany.

Pl. Syst. Evol., Suppl. 2, 73—88 (1979)

Institute of Biology II, Department of Genetics, The University
of Tübingen, German Federal Republic, and Department of
Botany, The University of Georgia, Athens, U.S.A.

Restriction Mapping of Ribosomal RNA Genes of
Higher Plants

By

Harald H. Friedrich, Vera Hemleben, Tübingen, and **Joe L. Key,** Athens,
U.S.A.

Key Words: Angiosperms, *Glycine max*, *Phaseolus aureus*, *Matthiola incana*,
Brassica pekinensis, *Cucumis melo.*—Ribosomal RNA genes, reatriction map.

Abstract: Soybean rDNA was enriched by preparative buoyant density
centrifugation in CsCl-actinomycin D gradients. The restriction endonuclease
mapping with *Bam* H I, *Hind* III, *Eco* R I, and *Bst* I demonstrated that there is
no heterogeneity, even in the spacer region of the soybean rDNA. It was possible
to arrange the fragments into a complete map. The repeating unit of the
ribosomal DNA has a molecular weight of 5.9×10^6 daltons. By hybridization
with ^{125}I-25 S and -18 S rRNA, the 25 S and 18 S rRNA coding sequences were
localized within the restriction map of the repeating unit. The arrangement of
the fragments and the apparent homogeneity of repeated and clustered rRNA
genes were further supported by preparative gel electrophoresis of *Bst* I
fragments. Another higher plant, *Phaseolus aureus*, which is closely related to
Glycine max (both belong to *Fabaceae-Phaseoleae*), was mapped by the same
restriction endonucleases. The rDNA of *Phaseolus* cross-hybridized to soybean
rRNA, but the fragment arrangement was found to be different from that of
soybean. The 18 S rRNA coding region, however, was similar in both plants. The
ribosomal genes of two members of the family *Cruciferae*, *Brassica pekinensis*,
and *Matthiola incana*, were also investigated by restriction endonuclease
fragment mapping. The only homology in fragment arrangement between these
four plants was found for the 18 S rRNA coding region. The size of the repeating
unit is 6.1×10^6 (*Phaselous*), 6.0×10^6 (*Matthiola*), and 6.2×10^6 daltons
(*Brassica*). The fifth plant investigated, *Cucumis melo*, (*Cucurbitaceae*) was
quite different with respect to the repeating unit which was found to be much
larger than in the other four plants (7.6×10^6 daltons).

The organization of repetitive genes coding for ribosomal 28 S and
18 S rRNA was analyzed in detail for animals such as *Xenopus* or
Drosophila (WELLAUER & DAWID 1977). In many aspects the ribosomal

genes of higher plants obviously are organized in a similar way. The 25 S and 18 S rRNA coding sequences are arranged in tandem. Transcription is performed in a common precursor molecule which is stepwisely processed to the final 18 S and 25 S rRNA. The size of the precursor-rRNA, the intermediates of processing, and the final rRNA products are smaller than those of animals (HADJIOLOV & NIKOLAEV 1976).

The number of ribosomal genes in higher plants varies enormously even between species of the same family (INGLE & al. 1975), resulting in a percentage of ribosomal DNA differing from about 0.05% to about 3-4% of total DNA. We investigated the organization of the five different plants: *Glycine max* (0.2% rDNA), *Phaseolus aureus* (about 3% rDNA), *Matthiola incana* (0.5% rDNA), *Brassica pekinensis* (2.1% rDNA), and *Cucumis melo* (about 0.5% rDNA). From these plants it was possible to isolate enough rDNA to analyze the ribosomal gene organization in detail. Three kinds of approaches are necessary for the characterization of such a specific DNA sequence:

1) Isolation and purification of specific DNA sequences from total DNA.

2) Isolation of pure and homogeneous RNA complementary to the DNA sequences.

3) Mapping the specific DNA sequence by restriction endonuclease analysis and identification of the coding regions.

Materials and Methods

Plant Material: *Matthiola incana* (L.) R. BR. and *Brassica pekinensis* (LOUR.) RUPR. (*Brassicaceae-Cruciferae*). *Phaseolus aureus* and *Glycine max* (L.) MERR. (*Fabaceae-Phaseoleae*), and *Cucumis melo* (*Cucurbitaceae*) were cultivated under sterile conditions. 6–8 day old seedlings were used for DNA extraction according to HEMLEBEN & al. (1975).

Isolation and Characterization of Soybean DNA. 50 g of soybean embryonic axes were imbibed in 200 ml of Tris-EDTA-buffer (100 mM Tris, 50 mM EDTA/pH 7.5) for 2 h. The DNA was isolated and purified as described by HEMLEBEN & al. (1975). The neutral CsCl gradients and the CsCl-actinomycin D gradients were prepared as described by HEMLEBEN & al. (1977). Saturation hybridization with soybean DNA on filters and ^{3}H-25 S and 18 S rRNA and gradient hybridization were performed as described by GRIERSON & HEMLEBEN (1977).

Isolation, Purification, and Iodination of 18 S and 25 S rRNA. Soybean seeds (*Glycine max* cv."Wayne"obtained from Noble Brothers Seed. Co., Gibson City, Illinois) were germinated, grown, and 2,4-dichlorphenoxyacetic acid-treated as described by GUILFOYLE & al. (1976). The hypocotyls were harvested, cut in small pieces of about 0.5 cm, placed in distilled water, and gased with nitrogen for 2 h to dissociate polysomes to monosomes (LIN & KEY 1971). The distilled water was poured off and replaced by extraction buffer (50 mM Tris-HCl, 10 mM

MgCl$_2$, 20 mM KCl, 0.01 % mercaptoethanol, 250 mM sucrose/pH 7.5). The hypocotyls were then homogenized with a Polytron for 60 sec at setting 8. The homogenate was centrifuged in a Sorvall SS 34 rotor at 9,000 rpm for 20 min; the supernatant was then centrifuged in a Beckman 60 Ti rotor at 45,000 rpm at 2 °C for 90 min. The sediment was suspended in 5 ml gradient buffer (50 mM Tris-HCl, 5 mM MgCl$_2$, 500 mM KCl/pH 7.5) and centrifuged again. The supernatants were combined and 75 A$_{260}$ units of ribosomes were layered on linear sucrose gradients (15-30 % or 15-33 %). The gradients were run in a Beckman SW 27 rotor at 20,000 rpm for 18 h at 20 °C. The gradients were fractionated by a flow-cell Gilford spectrophotometer arrangement. The fractions containing 40 S and 60 S subunits were each combined separately, phenol-extracted, and ethanol-precipitated. After reprecipitation the separated RNA species were dissolved in gradient buffer (100 mM NaCl, 10 mM Tris-HCl/pH 8.0) and run through an exponential 0.3 M to 1.4 M sucrose gradient in a Beckmann SW 41 rotor at 40,000 rpm for 16 h at 4 °C. The gradients were fractionated and the fractions of 5 S, 18 S, and 25 S were each combined separately, dialyzed overnight against gradient buffer and then ethanol-precipitated. The 18 S and 25 S rRNAs were dissolved and dialyzed against distilled water. The two rRNA species were iodinated using methods described elsewhere (GETZ & al. 1972, OREZ & WERTMUR 1974).

Digestion of Ribosomal DNA. The enriched rDNA obtained by CsCl-actinomycin D gradients was digested with *Eco* R I, *Bam* H I, and *Hind* III restriction endonucleases under the following incubation conditions:
Eco R I (Miles Laboratories Inc. Research products):
10 mM Tris-HCl, 50 mM NaCl, 10 mM MgCl$_2$/pH 7.5, 37 °C, 2 h.
Bam H I (New England Biolabs):
50 mM NaCl, 6 mM MgCl$_2$, 6 mM Tris-HCl, 6 mM mercaptoethanol, 10 % bovine serum albumin/pH 7.4, 37 °C, 2 h.
Hind III (New England Biolabs):
60 mM NaCl, 7 mM MgCl$_2$, 7 mM Tris-HCl/pH 7.4, 37 °C, 2 h.
Each incubation mixture had a volume of 50 μl and contained 2 μg of rDNA and 10 units of restriction enzyme. The reaction was stopped by adding 5 μl of stop mix (5% SDS, 50% glycerol, 0,025% bromphenol blue).

Electrophoresis of the Digested rDNA. The agarose slab gels were prepared between 2 glass plates. At the bottom of the gel apparatus a 2-3 cm long polyacrylamide gel was polymerized. The agarose solution (1 %, 1.5 %, or 2 % w/v) was boiled in 1 × electrophoresis buffer (36 mM Tris, 30 mM NaH$_2$PO$_4$, 2 mM EDTA/pH 7.7), then was cooled down to 65 °C and poured in the gel apparatus to a size of 17 × 16 × 0.3 cm. The digested DNA samples were placed onto the gel and after a start electrophoresis of 150 mA for 10 min, the gel was run overnight at 40 V circulating the electrophoresis buffer. *Eco* R I and *Hind* III digested λ DNA were used as makers.
After electrophoresis the gel was stained for 45 min with ethidium bromide (20 μg/ml in 10 × electrophoresis buffer), photographed, and cut in strips. The denatured DNA was eluted from the agarose onto a nitrocellulose filter and then hybridized to ^{125}I-25 S and ^{125}I-18 S rRNA or ^{125}I-25 S, 18 S, 5 S rRNA alone (in the presence of fifty-fold excess of the corresponding unlabelled rRNA as competitor) in 6 × SSC and 25 % formamide according to methods described by SOUTHERN (1975). The hybridized filter strips were exposed to a Kodak no screen X-ray film.

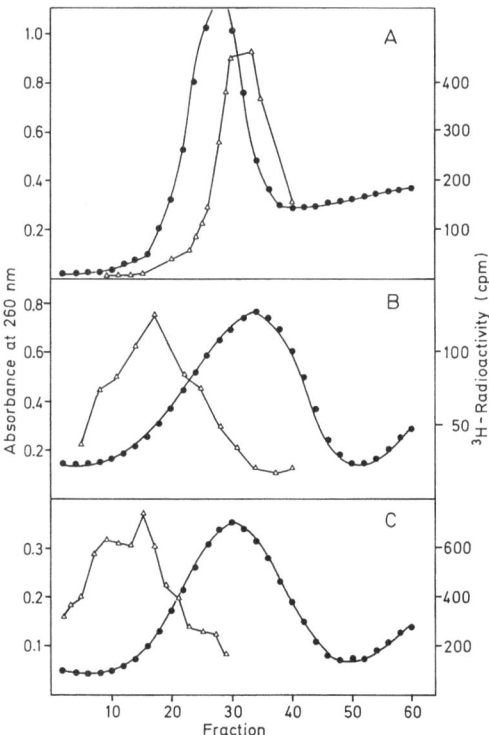

Fig. 1. Enrichment of soybean ribosomal DNA sequences. — A. CsCl-gradient:
600 µg soybean DNA were loaded on neutral CsCl gradients and centrifuged in a
Beckman 50 Ti rotor at 32,500 rpm and 2 °C for 60 h. 50 µl of each 250 µl-fraction
were taken and the DNA from these samples was denatured, fixed onto nitrocellulose filters and then hybridized to ^{3}H-18S and 25S rRNA (2 × SSC; 70°C; 3h). The
filters were washed, RNAse treated, and counted. — B. First CsCl-actinomycin
D gradient: 250 µg rDNA from the neutral CsCl gradients (A) identified by
gradient hybridization were loaded on CsCl-actinomycin D gradients and
centrifuged in a Beckman 50 Ti rotor at 30,000 rpm and 3 °C for 84 h. 20 µl of
each fraction were taken and gradient hybridization was performed. — C.
Second CsCl-actinomycin D gradient: 125 µg enriched rDNA from the CsCl-
actinomycin D gradients (B) were loaded on CsCl-actinomycin D gradients and
centrifuged. 10 µl of each fraction were used for hybridization. (●—●—●)
Absorbance at 260 nm; (△—△—△) Radioactivity (^{3}H-rRNA/rDNA hybrid)

Preparative Agarose Gel. 8 mg total soybean DNA was diluted to a final
concentration of 0.1 mg/ml in 100 mM Tris-HCl, 5 mM MgCl$_2$/pH 7.5. The
solution was incubated for 40 min at 37 °C with approximately 10,000 units of
Bst I prepared according to the method of CATTERALL & WELKER (1977). The
digested DNA was phenol-extracted and then four times ether-extracted. The
aqueous phase was ethanol-precipitated overnight and then redissolved in 10 ml

electrophoresis buffer. This sample was placed onto a 1% preparative cylindrical agarose gel (CAREIRA & MEAGHER, in press). The electrophoresis was run at a constant voltage of 30 V and 140 mA about 17 h until the bromphenol blue dye had reached the inner center of the gel. Then the gel was run for a further 52 h, and at 20 min intervals 6 ml-fractions were collected. An aliquot was taken from each fraction, the DNA denatured by the addition of an

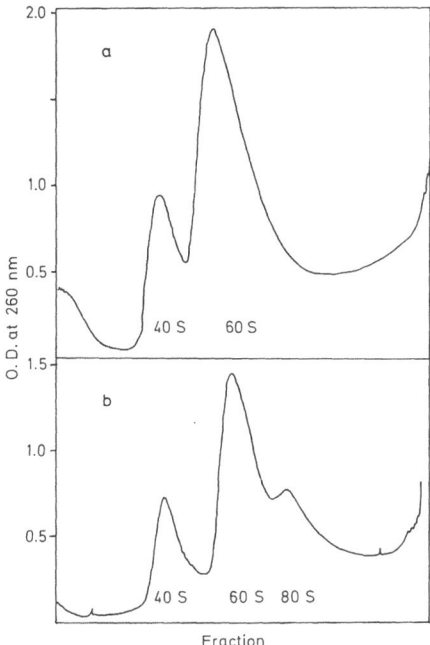

Fig. 2. Separation of soybean ribosomal subunits. — The suspensions (see Materials and Methods) were layered on linear sucrose gradients (a: 15-30% / b: 15-35%) and centrifuged at 20,000 rpm, 18 h, and 2 °C with a Beckman SW 27 rotor. (————) Absorbance at 260 nm

equal volume 100 mM NaOH, neutralized, and loaded onto nitrocellulose filters (24 mm diameter). The filters were placed into [125]I-25 S and [125]I-18 S rRNA solution for DNA-rRNA hybridization (6 × SSC, 25% formamide, 65 °C, 18 h).

Results and Discussion

In this work we used the method of CsCl-actinomycin D buoyant density centrifugation (HEMLEBEN & al. 1977) for enrichment of ribosomal DNA. Actinomycin is intercalating between GC-pairs in the DNA double helix. DNA sequences which are GC-rich like rDNA are

H. H. Friedrich et al.:

shifted from the main peak to a lower buoyant density. In Fig. 1 A a
normal neutral CsCl buoyant density gradient of soybean (*Glycine max*)
DNA is shown. From each fraction an aliquot was tested by filter
hybridization with ³H-rRNA from *Matthiola incana*. The fractions from

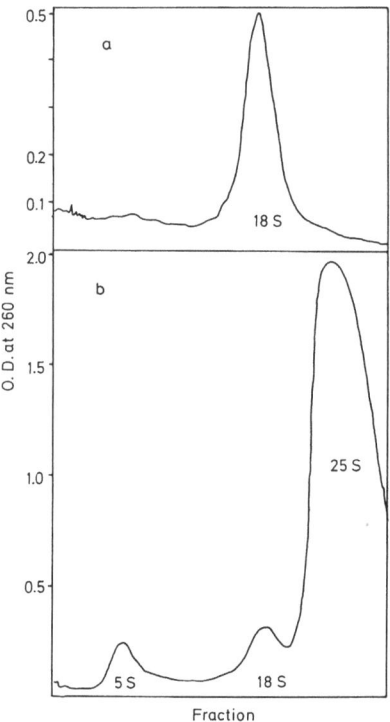

Fig. 3. Characterization of the RNA from soybean ribosomal subunits. — The
rRNA was isolated from the ribosomal subunits as described in Materials and
Methods and then layered on isokinetic sucrose gradients: a) rRNA isolated from
the 40 S ribosomal subunit; b) rRNA isolated from the 60 S ribosomal subunit.
The gradients were centrifuged at 40,000 rpm, 16 h, and 4 °C with a Beckman
SW 41 rotor

that gradient containing rDNA were pooled and recentrifuged on a CsCl-
actinomycin D gradient. Filter hybridizations were performed again
(Fig. 1 B). A comparison of both figures demonstrates the separating effect
of actinomycin D on rDNA. The rDNA containing fractions from that
gradient were pooled again and recentrifuged on CsCl-actinomycin D
gradients (Fig. 1 C). The enrichment after three gradients is about 65-70-
fold. This enriched rDNA was used for the following experiments.

The second step was the purification of the 18 S and 25 S rRNA. We first isolated ribosomes from soybean hypocotyls, and then separated the two subunits after dissociation on linear sucrose gradients (Fig. 2 a, b) (LIN & KEY 1971). It is important to form gradients steep enough to

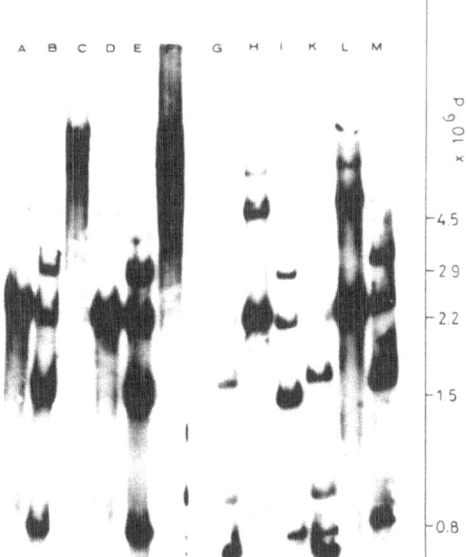

Fig. 4. Restriction endonuclease digestion pattern of enriched soybean rDNA. — 2 μg of enriched soybean rDNA were each incubated with *Eco* R I (A and D), *Bam* H I (B and E), and *Hind* III (C and F) restriction endonucleases or for a double digest with two of the enzymes: *Bam* H I followed by *Eco* R I (G and K), *Hind* III followed by *Eco* R I (H and I), and *Bam* H I followed by *Hind* III (I and M). The digested DNA samples were separated on an 1 % agarose gel, eluted onto nitrocellulose strips, and hybridized to 125I-18 S and -25 S soybean rRNA as described in Materials and Methods. After hybridization the strips were exposed to a Kodak no screen X-ray film for 10 days

move a possibly undissociated contamination out of the 60 S subunit band. Fig. 2 a and b show differences in separation between a linear sucrose gradient from 15–30% and 15–33% sucrose. The RNA was isolated from the separately pooled subunits and then fractionated on isokinetic sucrose gradients (Fig. 3 a, b). We received pure 18 S rRNA from the 40 S subunit and nearly pure 5 S and 25 S rRNA with a small contamination of 18 S rRNA from the 60 S subunit. The contamination of 18 S rRNA is related to the contamination of undissociated 80 S monosomes in the 60 S subunit peak. The fractions for each rRNA were

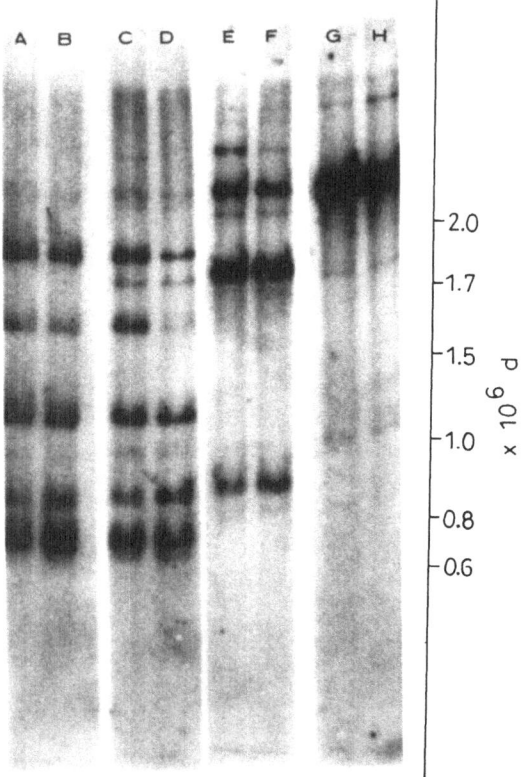

Fig. 5. Restriction endonuclease digestion of enriched soybean rDNA. — 2 µg of
enriched soybean rDNA were each incubated with *Bam* H I followed by *Eco* R I
restriction endonuclease (A-D), *Hind* III followed by *Bam* H I restriction
endonuclease (E and F), and *Hind* III followed by *Eco* R I restriction
endonuclease (G and H). The digested DNA samples were separated on a 2%
agarose gel, eluted, and hybridized as described for Fig. 4 and in Materials and
Methods

pooled separately and iodinated *in vitro* to get labelled rRNA with high
specific radioactivity.

The enriched rDNA was digested with different restriction
endonucleases, separately and in combination; the digested rDNA
fragments were separated on agarose gels. After electrophoresis the
bands were eluted on nitrocellulose filter strips (SOUTHERN 1975) and
finally hybridized to ^{125}I-rRNA. The results of the restriction
endonuclease digestion fragment mapping is shown in Figs. 4 and 5, and
Table 1. The restriction endonuclease *Bam* H I leads to three complete

digested DNA fragments hybridizing with ^{125}I-soybean rRNA: 2.2, 1.5 and 0.8 × 10⁶ daltons. Incomplete digested fragments with the molecular weight of 2.9 × 10⁶ daltons (2.2 + 0.8 × 10⁶ daltons), 3.8 × 10⁶ daltons (2.2 + 1.5 × 10⁶ daltons) and 4.5 × 10⁶ daltons (2.2 + 1.5 + 0.8 × 10⁶ daltons) were also obtained. These results lead to a *Bam* H I fragment map of soybean rDNA (Fig. 6 A).

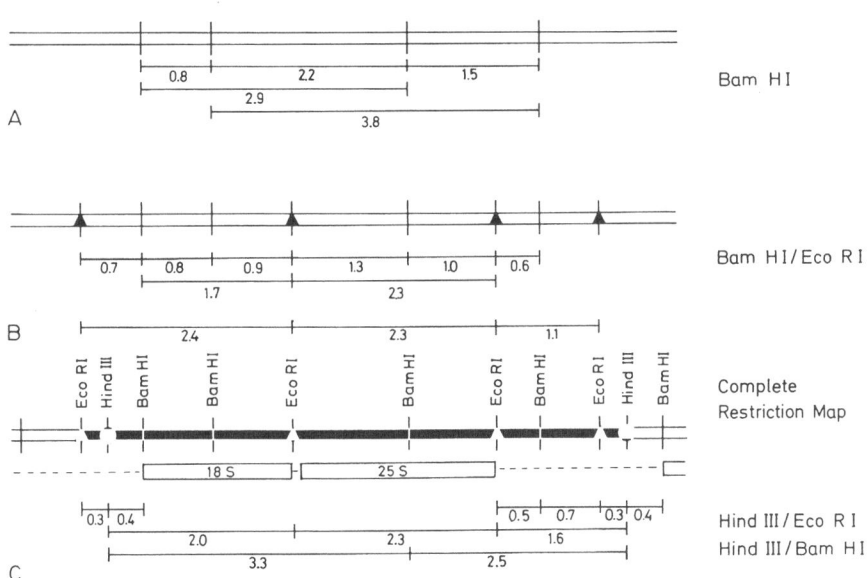

Fig. 6. Restriction map of soybean ribosomal genes. — A. *Bam* H I restriction map; B. *Bam* H I / *Eco* R I restriction map; C. complete restriction map with the recognition sites for *Bam* H I, *Eco* R I, and *Hind* III and localization of the 18 S and 25 S rRNA coding sequences. All numbers are abbreviations of the molecular weight in 10⁶ daltons

The restriction endonuclease *Eco* R I produced three bands: 4.7 (very weak hybridization), 2.3 2.45 and 1.1 × 10⁶ daltons. A mapping analysis was not possible by these few data. Therefore a double digest with *Bam* H I followed by *Eco* R I was performed and the following rDNA fragments hybridizing with rDNA were received (Table 2). In addition to the original *Bam* H I fragments of 2.2, 1.5, 0.8 × 10⁶ daltons and the original *Eco* R I fragments of 2.3 and 1.1 × 10⁶ daltons, five bands with the molecular weight of 1.7, 1.3, 1.0, 0.9, and 0.6 × 10⁶ daltons were received.

82   H. H. Friedrich et al.:

Table 1. Fragments resulting from restriction endonuclease single digests of
soybean rDNA

Enzyme	Molecular weight of the fragments
Bam H I	4.5×10^6
	3.8×10^6
	2.9×10^6
	2.2×10^6
	1.5×10^6
	0.8×10^6
Eco R I	4.7×10^6
	$2.3\text{–}2.45 \times 10^6$
	1.1×10^6
Hind III	5.9×10^6 (in a smear from 5 to 12.5×10^6)

Table 2. Fragments resulting from restriction endonuclease double digest of
soybean rDNA

Enzyme I	Enzyme II	Molecular weight of the fragments ($\times 10^6$ daltons) hybridizing to ^{125}I-18 S and -25 S soybean rRNA	Specific hybridization to .... rRNA
Bam H 1	*Eco* R 1	2.9	25 S
		2.3	18 S
		2.2	25 S
		1.7	18 S
		1.5	25 S
		1.3	25 S
		0.9	18 S
		0.8	18 S
		0.6	—
Hind III	*Bam* H I	5.9	
		4.5	
		3.3	
		2.5	
		2.2	
		1.5	
		0.8	
Hind III	*Eco* R I	10.3	—
		7.9	—
		5.9	25 S
		2.3	25 S
		2.0	—
		1.6	—

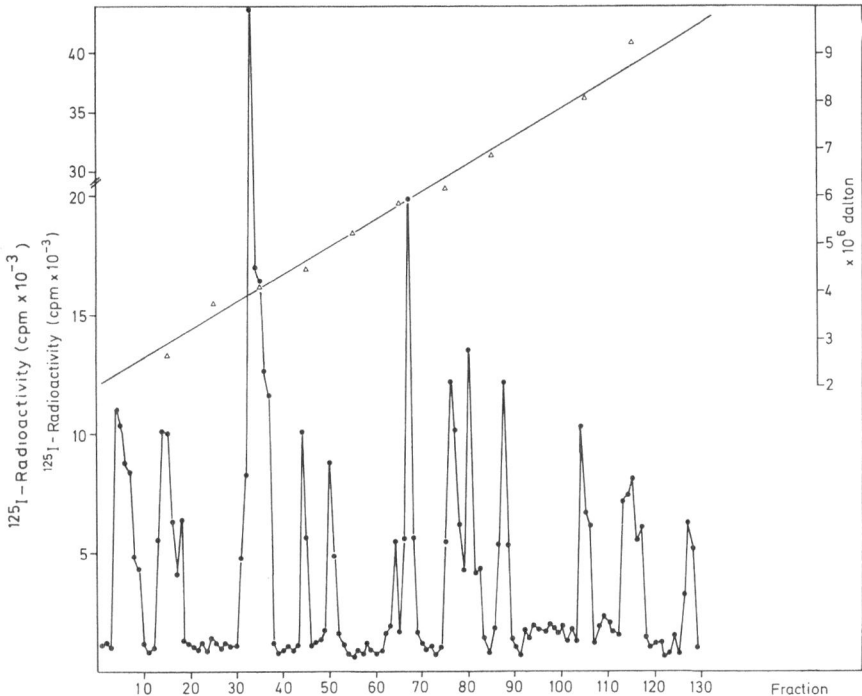

Fig. 7. Preparative gel electrophoresis of *Bst* I fragments. — 8 mg total soybean DNA were incubated with *Bst* I restriction endonuclease under incomplete digestion conditions and then phenol-extracted. The digested soybean DNA was loaded onto a 1 % preparative agarose gel. Electrophoresis, fractionation, and hybridization with ^{125}I-18 S and -25 S soybean rRNA of each fraction were performed as described in Materials and Methods. An aliquot of every 10th fraction was run again on a 1% agrose slab gel to determine the molecular weight △—△—△   (●—●—●) ^{125}I-radioactivity (^{125}I-rRNA/rDNA hybrids)

The fragments of 1.3 and 1.0 × 10⁶ daltons are parts of the original 2.3 × 10⁶ dalton *Eco* R I band, which means that one *Bam* H I recognition site is located inside that *Eco* R I fragment. The fragments of 0.9 and 0.6 × 10⁶ daltons are parts of the original 1.5 × 10⁶ *Bam* H I fragment, which therefore shows one *Eco* R I recognition site inside this *Bam* H I fragment. The 1.7 × 10⁶ dalton fragment consists of the 0.8 × 10⁶ dalton original *Bam* H I fragment and the 0.9 × 10⁶ dalton *Bam* H I/*Eco* R I double digest band which is cut by *Eco* R I out of the original *Bam* H I fragment of 2.2 × 10⁶ daltons. Therefore, a *Bam*

6*

Table 3. Comparison of the molecular weights of the Bst I fragments received by preparative gel electrophoresis (see Fig. 7) with the theoretical fragment sizes derived from the restriction map (see Fig. 6)

Result molecular weight of the obtained fragments (see Fig. 7) ($\times 10^6$ daltons)	*Bam* H I fragment arrangement								Theoretical molecular weight of the fragments ($\times 10^6$ daltons)
	1.4	0.8	2.2	1.5	1.4	0.8	2.2	1.5	
5.0									5.1
5.9									5.9
6.5									
6.7									6.7
7.2									7.3
8.1									8.1
8.8									8.8
9.6									9.6

H I/*Eco* R I double digest fragment map could be established and derived from the *Eco* R I fragment map (Fig. 6 B).

After digestion with the *Hind* III restriction endonuclease only a smear in the range of 12.5-5×10^6 daltons with a band of 5.9×10^6 daltons could be received. For further details double digests with *Bam* H I and *Eco* R I were performed. The results of these double digests are summarized in Table 2. The *Hind* III/*Bam* H I double digests demonstrate that the *Hind* III recognition site is located outside the 4.5×10^6 dalton *Bam* H I region. The *Hind* III/*Bam* H I double digest fragments of 3.3 and 2.5×10^6 daltons help to determine the distance of the *Hind* III recognition site from both ends of the 4.5×10^6 dalton fragment. The *Hind* III recognition site is located outside the DNA sequences coding for 25 S and 18 S rRNA. The *Hind* III/*Eco* R I double digestion verifies the localization of the *Hind* III sites within the rDNA sequences (Fig. 6).

These data suggest that the repeating unit of soybean rDNA is 5.9×10^6 daltons, and *Hind* III cuts it only once per unit. Preparative agarose gel electrophoresis of a large amount of soybean DNA digested by *Bst* I restriction endonuclease (same recognition site as *Bam* H I) was performed under limiting time conditions, and hybridization of each fraction to ^{125}I-rRNA, in order to confirm the repeating unit length, and to learn whether the repetitive ribosomal genes are homogeneously or heterogeneously arranged. Limiting incubation times lead to incomplete digestions. Therefore, rDNA fragments with a molecular weight larger than one repeating unit should be received. If the length of the repeating unit is constant, only fragments with a molecular weight of additive size of the complete *Bst* I fragments can be expected. Fig. 7 shows the molecular weight pattern after preparative gel electrophoresis of the *Bst* I fragments. In Table 3 the actually found molecular weights of the

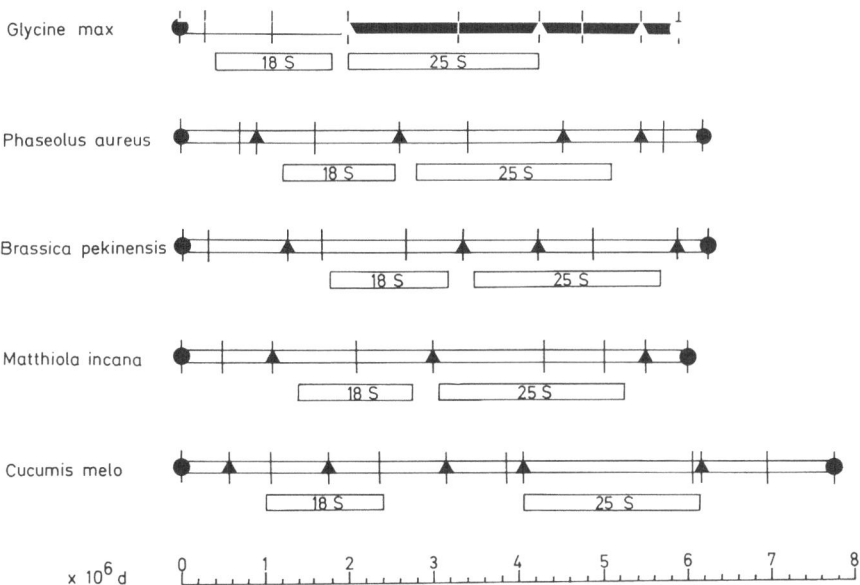

Fig. 8. Restriction map for several higher plants. — Complete restriction map with the recognition sites for *Bam* H I, *Eco* R I, and *Hind* III and localization of the 18 S and 25 S rRNA coding sequences

fragments are compared with the calculated theoretical lengths: There is nearly no difference between both molecular weights. This result shows that there is a high homogeneity in the length of the repeating units.

The next problem was to localize the coding sequences within the repeating unit. "SOUTHERN hybridizations" were performed under competition hybridization conditions: [125]I-18 S rRNA was hybridized to digested soybean rDNA with an excess of a 50-fold unlabelled 25 S rRNA and [125]I-25 S rRNA was hybridized in the presence of a 50-fold unlabelled 18 S rRNA. The results are shown in Fig. 6 C and Table 2. From the end of a 25 S rRNA coding sequence in one repeating unit to the beginning of the 18 S rRNA coding sequence in the following repeating unit, there is a distance in length of 2.0×10^6 daltons. This spacer still contains sequences coding for the rRNA precursor which is processed during maturation of ribosomal RNA. Taking the molecular weight of the precursor as 2.2×10^6 daltons (GRIERSON & HEMLEBEN 1977) leaves a nontranscribed spacer of about 1.4-1.5×10^6 daltons between two repeating units.

The sequence organization found for the ribosomal genes of soybean (*Glycine max*) then was compared with those of other higher plants. The rDNA of *Phaseolus aureus, Brassica pekinensis, Matthiola incana* (both crucifers), and *Cucumis melo* was isolated and enriched as described for soybean. Restriction endonuclease digestion, gel electrophoresis and "SOUTHERN hybridization" were performed in the same way. The mapping data are summarized in Fig. 8. The ribosomal gene map of *Phaseolus aureus*, which is closely related to *Glycine max* (both belong to the *Fabaceae-Phaseoleae*), shows only some homology in the fragment pattern to soybean ribosomal RNA genes. The size of the repeating unit is 6.1×10^6 d, i.e. in the range of that found for soybean. The sizes of single and double digest fragments from *Phaseolus* are in some cases comparable to soybean fragments, but the arrangement of these fragments to the complete repeating unit is different. Comparing the localization of the coding sequences within the repeating unit demonstrates some similarities for the 18 S coding sequence. The 0.9×10^6 dalton *Bam* H I/*Eco* R I double digest fragment seems to be homologous in both plants, whereas the 25 S region seems to have some sequence alterations in comparison to soybean.

From the plant family *Cruciferae* (= *Brassicaceae*) two members were investigated: *Brassica pekinensis* and *Matthiola incana* (Fig. 8). The size of the rDNA repeating unit from both plants is in the range of the plants investigated before (6.2 and 6.0×10^6 d). The results from single and double digestions are quite different: *Brassica pekinensis* contains four *Eco* R I sites (as *Phaseolus*) in comparison to three *Eco* R I sites in *Matthiola* and *Glycine max*. For the 18 S rRNA coding sequence there is the same fragment homology for the 0.9×10^6 dalton *Bam* H I/*Eco* R I double digest fragment, whereas there is no fragment similarity for the 25 S rRNA coding region.

The fifth plant investigated was *Cucumis melo*, a member of the *Cucurbitaceae*. It has a very large repeating unit of 7.6×10^6 daltons in comparison to the other four plants. The coding sequences for 18 S and 25 S rRNA are separated from one another by a larger spacer sequence (Fig. 8). Even in the region of the coding sequences a very different fragment pattern was found. This result is in agreement with other special features of *Cucumis melo* in respect to the organization of the ribosomal genes (HEMLEBEN & GRIERSON 1978).

The comparison of the restriction maps of the ribosomal genes of various higher plants demonstrates that there are structural similarities. Sequence homology of the ribosomal RNA coding region between different higher plants is shown by a high amount of cross hybridisation. The differences in specific restriction endonuclease sites demonstrate that base exchanges occured as well in the transcribed region as in the nontranscribed spacer sequences of the ribosomal DNA repeating units. The 18 S rDNA sequences are obviously more conservative than the 25 S rDNA sequences. The fact that each repeating unit of all plants investigated so far contains only one *Hind* III recognition site offers a good possibility to isolate a complete specific gene.

The research on mapping soybean rDNA was supported by Public Health Research Ca 11624 from the National Cancer Institute to Joe L. Key. The authors are greatly indebted to Dr. J. R. Y. Rawson for his helpful comments. Part of this work was supported by the *Deutsche Forschungsgemeinschaft*.

References

Catterall, J., Welker, N., 1977: Isolation and properties of a thermostable restriction endonuclease (Endo R *Bst* 1503). — Bact. **129**, 1110—1120.

Getz, M. J., Altenburg, L. C., Saunders, G. F., 1972: The use of RNA labelled *in vitro* with iodine-125 in molecular hybridization experiments. — Biochim. Biophys. Acta **287**, 485—494.

Grierson, D., Hemleben, V., 1977: Ribonucleic acid from the higher plant *Matthiola incana* / Molecular weight measurements and DNA, RNA hybridization studies. — Biochim. Biophys. Acta **475**, 424—436.

Hadjiolov, A. A., Nikolaev, N., 1976: Maturation of ribosomal ribonucleic acids and the biogenesis of ribosomes. — Prog. Biophys. Molec. Biol. **31**, 95—144.

Hemleben, V., Ermisch, N., Kimmich, D., Leber, B., Peter, G., 1975: Studies on the fate of homologous DNA applied to seedlings of *Matthiola incana*. — Eur. J. Biochem. **56**, 403—411.

— Grierson, D., Dertmann, H., 1977: The use of equilibrium centrifugation in actinomycin-caesium chloride for the purification of ribosomal DNA. — Pl. Sci. Lett. **9**, 129—135.

— — 1978: Evidence that in higher plants the 25 S and 18 S rRNA genes are not interspersed with genes for 5 S rRNA. — Chromosoma (Berl.). **65**, 353—358.

Ingle, I., Timmis, I. N., Sinclair, J., 1975: The relationship between satellite desoxyribonucleic acid, ribosomal ribonucleic acid gene redundancy and genome size in plants. — Pl. Physiol. **55**, 496—501.

Lin, C. Y., Key, J. L., 1971: Dissociation of N_2 gas-induced monomeric ribosomes and functioning of the derived subunits in protein synthesis in pea. — Pl. Physiol. **48**, 547—552.

Oroz, I. M., Wertmur, J. B., 1974: *In vitro* iodination of DNA: Maximizing iodination while minimizing degradation; use of buoyant density shifts for DNA—DNA hybrid isolation. — Biochemistry **13**, 5467—5473.

Southern, E. M., 1975: Detection of specific sequences among DNA fragments separated by gel electrophoresis. — J. Molec. Biol. **98**, 503—517.

WELLAUER, P. K., DAWID, I. B., 1977: Organization of members within the repeating families of the genes coding for ribosomal RNA in *Xenopus laevis* and *Drosophila melanogaster*. In BEERS, R. F., BASSETT, E. G. (Eds.): Recombinant Molecules: Impact of Science and Society. New York: Raven Press.

Address of the authors: Dipl.-Biol. HARALD H. FRIEDRICH and Prof. VERA HEMLEBEN, Institut für Biologie II, Lehrstuhl für Genetik der Universität Tübingen, Auf der Morgenstelle 28, D-7400 Tübingen, Federal Republic of Germany.

Differential DNA Replication

Pl. Syst. Evol., Suppl. 2, 91—94 (1979)

Institute of Developmental Physiology,
University of Cologne, Federal Republic of Germany

Cytophotometric Measurements of DNA in Salivary Gland Nuclei of *Drosophila melanogaster*

By

Leonore Dennhöfer, Cologne

Key Words: *Drosophila.* — DNA, cytophotometric measurement, polytene nuclei.

Abstract: Comparative measurements of nuclear DNA (Feulgen) after hydrolysis with 1 n HCl at 60 °C and 5 n HCl at 20 °C revealed that only the latter gives accurate results. Therefore, the problem of heterochromatin underreplication in *Drosophila* polytene chromosomes should be re-investigated using coldhydrolysed material, because older data may be the result of a nonsuitable method and, therefore, erroneous.

In order to find a solid basis for further investigations, we looked for the maximum size of larval salivary gland nuclei *in vivo*. Cytophotometric measurements of the Feulgen-stained nuclei point out that 4-hours old female prepupae have reached in about 6 % of the nuclei the maximum size of 4096 C (= 2 n^{12}). The DNA classes are established on the basis of 4 C values obtained in ganglia cells and cells of the imaginal salivary gland anlage. This value is in good agreement with the 2 C value in these tissues, and the C value of spermatocytes (DENNHÖFER 1979). Studies on polyploid series of nuclei require a method, which allows to measure the DNA content in small nuclei as exactly as in big nuclei, when lying together in the same tissue.

Methods

Salivary glands were squashed, and fixed according to LILLIE (ethanol:formol:acetic acid = 85:10:5), and the Feulgen reaction was performed according to GRAUMANN (1953). The salient point, however, is the hydrolysis of DNA before the Feulgen reaction. It is known, that an optimum cytophotometric measurement of DNA is only possible, if the hydrolysis of the DNA is standardized and is in the optimum extent. In most laboratories, the so-called hot hydrolysis (1 n HCl, 60 °C) is performed. Using this method, each nucleus, i.e. each DNA content, shows the maximum of hydrolysed purines at a different time of treatment. Thus, the same time may be right for one nucleus, but too short for other nuclei, so that the extinction values do not correspond to the DNA contents. When the duration of hydrolysis exceeds the optimum, the DNA is dissolved by HCl, i.e. the cytophotometric value is too low again. Therefore, hot hydrolysis is not useful for tissues with nuclei of different sizes.

SANDRITTER & al. (1964, 1965), Fox (1969) and VAHS (1972) described two methods of hydrolysis especially for tissues with nuclei of different ploidy levels. We introduced the method of cold hydrolysis according to VAHS (1972) to our object (5 n HCl, 20 °C, 30-50 min).

Results

The differences of hot and cold hydrolysis are shown by comparing cytophotometric values of small and big nuclei in Table 1. For all experiments, larval salivary glands, together with the attached imaginal salivary gland anlage, were taken from larvae just after puparium formation. In this developmental stage most of the salivary gland nuclei belong to 2048 C or to the neighbouring small intermediate size group, while the nuclei from the salivary gland anlage cells are just in the running duplication step from 2 C to 4 C (DENNHÖFER 1979).

Table 1. Measurements of DNA in Feulgen-stained nuclei in salivary glands from 0-hour-old prepupae after different methods of hydrolysis. The extinction integrals are given in arbitrary units.

Method of hydrolysis	Nuclei from imaginal salivary gland anlage (number of nuclei in parentheses)		Maximum size of salivary gland nuclei from 0-h-old prepupae	
	♀	♂	♀	♂
Hot hydrolysis 8 min	—	3,39 (10)	2204	1910
Hot hydrolysis 10 min	3,06 (15)	3,17 (4)	2226	2045
Hot hydrolysis 12 min	0	0	2305	2204
Hot hydrolysis 14 min	0	0	2509	—
Hot hydrolysis 16 min	0	0	2271	2192
Cold hydrolysis 40 min	3,36 (272)	3,28 (308)	3718	3768

The cytophotometric values of small nuclei from the anlage cells were the same after hot hydrolysis of 8 min and after cold hydrolysis, the big salivary gland nuclei, however, did not reach the values as obtained after cold hydrolysis. After 10 min of hot hydrolysis the value of the small nuclei have decreased, i.e. hydrolysed components of the DNA were already dissolved, while the value of the big nuclei have still increased. After hot hydrolysis for longer than 12 min, the imaginal salivary gland anlage contained less nuclei than usual; this indicates that the DNA of small nuclei was completely dissolved. The big nuclei show high values, but they never reach the same extinction integrals as after cold hydrolysis. After hot hydrolysis for 16 min or more, also the cytophotometric values of the big salivary gland nuclei have decreased due to DNA depolymerization.

Discussion

This comparative studies demonstrate that only after cold hydrolysis Feulgen-stained nuclei may give optimum cytophotometric values, particularly of small nuclei (2C-4C) and polyploid nuclei (2048 C), lying close together in the same tissue. Evidently this is, because cold hydrolysis leads to a stable plateau of hydrolysed components of DNA for a rather long time (VAHS 1972), i.e. cold hydrolysis allows the DNA of big nuclei to be hydrolysed completely, while the DNA of the small nuclei does not become degraded during that time.

RODMAN (1967) has found a maximum value of only 256 C ($= 2\,n^8$) in the salivary gland nuclei of *Drosophila melanogaster*. RUDKIN (1969) put forward the hypothesis of non-replicating DNA in polytene chromosomes of nuclei from male larvae, due to the heterochromatic Y-chromosome, because the Feulgen-extinction values of these chromosomes did not follow a geometrical order in relation to the 2 C value. As the nuclear DNA content in nuclei of both sexes are multiples of the 4 C value after cold hydrolysis, as was shown by DENNHÖFER (1979), we suggest a re-investigation of the problem of DNA underreplication in *Drosophila*. The results by RUDKIN (1969) may be erroneous, because he used an unfit method before cytophotometric DNA measurement. On the other hand, biochemical studies in *Drosophila* (e.g. ENDOW & GALL 1975) and cytological data from plant material, which was Feulgen-stained after cold hydrolysis, are in agreement with the hypothesis of DNA underreplication (reviewed by NAGL 1978; see also this volume, e.g. SCHÄFFNER & NAGL 1979).

This work has been supported by the Deutsche Forschungsgemeinschaft.

References

DENNHÖFER, L., 1979: Growth of polytenic chromosomes in the nuclei of prepupal salivary glands in *Drosophila melanogaster*. — (In preparation).
ENDOW, S. A., GALL, J. G., 1975: Differential replication of satellite DNA in polyploid tissues of *Drosophila virilis*. — Chromosoma **50**, 175—192.

FOX, D. P., 1969: Some characteristics of the cold hydrolysis technique for staining plant tissues by the Feulgen reaction. — J. Histochem. Cytochem. **17**, 266—272.

GRAUMANN, W., 1953: Zur Standardisierung des Schiffschen Reagens. — Zeitschr. Wiss. Mikroskopie **61**, 225—226.

NAGL, W., 1978: Endopolyploidy and Polyteny in Differentiation and Evolution. — Amsterdam: North-Holland.

RODMAN, T. C., 1967: DNA replication in salivary gland nuclei of *Drosophila melanogaster* at successive larval and prepupal stages. — Genetics **55**, 375—386.

RUDKIN, G. T., 1969: Non replicating DNA in *Drosophila*. — Genetics **61**, Suppl. 1, 227—238.

SANDRITTER, W., BOSSELMANN, K., RAKOW, L., JOBST, K., 1964: Untersuchungen zur Feulgenreaktion. Die Langzeithydrolyse bei verschiedenen Zelltypen. — Biochim. Biophys. Acta **91**, 645—647.

— JOBST, K., RAKOW, L., BOSSELMANN, K., 1965: Zur Kinetik der Feulgenreaktion bei verlängerter Hydrolysezeit. Cytophotometrische Messungen im sichtbaren und ultravioletten Licht. — Histochemie **4**, 420—437.

SCHÄFFNER, K.-H., NAGL, W., 1979: Differential DNA replication involved in transition from juvenile to adult phase in *Hedera helix*. — Pl. Syst. Evol., Suppl. **2**, 105—110.

VAHS, W., 1972: Die Bedeutung der Hydrolyse-Art in der Feulgen-Cytophotometrie von Kernen mit unterschiedlichen Ploidiegraden. — Histochemie **33**, 341—348.

Address of the author: Dr. LEONORE DENNHÖFER, Institut für Entwicklungsphysiologie der Universität zu Köln, Gyrhofstraße 17, D-5000 Köln 41 (Lindenthal), Federal Republic of Germany.

Pl. Syst. Evol., Suppl. 2, 95—103 (1979)
© by Springer-Verlag 1979

Institute of Plant Nutrition, Department of Tissue Culture,
University of Giessen, Federal Republic of Germany

Qualitative Differences in the DNA of Some Higher Plants, and Aspects of Selective DNA Replication During Differentiation

By

Elke Dührssen, Angelika Schäfer, and Karl-Hermann Neumann,
Giessen

Key Words: Angiosperms, *Daucus carota*, *Petroselinum crispum*, *Datura innoxia*. — Differentiation, selective DNA replication.

Abstract: The kinetic complexity of the genome of *Daucus carota* was studied by reassociation experiments and compared with that of *Petroselinum* and *Datura innoxia*, and about 20% homology were found by cross-hybridization studies. Small differences in cot fractions occur between certain varieties of carrot. Renaturation experiments and density gradient studies indicated that carrot tissues display various fractions of repetitive DNA and satellite DNA, respectively, at differing developmental stages and after application of gibberellic acid. While these results indicate the selective replication of certain DNA sequences during development, microfluorometric DNA measurements do not. Therefore, we have obtained some evidence, but no direct proof, for differential DNA replication in carrot tissue cultures.

Although many homologous events occur during ontogenesis of higher plants like the formation of roots, shoots, flowers, stomata etc., qualitative as well as quantitative differences are apparent at all levels of organization, as expressed by the vast number of species classified as higher plants. The speciesspecific genome is transmitted from one generation to the next, usually more or less unchanged, and most of its expression is subjected to classical genetic laws and manipulations. Therefore, both the similarities and differences between different species are usually considered as firmly fixed in the genetic material, i.e. in the nucleotide sequences and their organization in the DNA of chromosomes.

During the last decade, however, an increasing number of publications (for reviews: NAGL 1976, BUIATTI 1977) indicated changes in the composition of nuclear DNA during differentiation. This is in disagreement with the central dogma of identity of genetic information in all cells of an organism as most convincingly demonstrated by embryogenesis in e.g. carrot cell and tissue cultures of various somatic origin. A series of experiments was carried out, in order to obtain some informations on this rather paradox and complex situation: Firstly, we characterized the DNA of the carrot *Daucus carota* L., as related to the DNA of other species like parsley, *Petroselinum crispum* (MILL.) A. W. HILL, or *Datura innoxia*, MILL. and secondly we characterized the DNA of tissues of carrot at various developmental stages as influenced by phytohormones like kinetin or gibberellic acid. These investigations are far away from being complete, and in the following only some first results will be discussed. One major problem resides in the fact that genetic changes *within* a species (i.e. between different varieties or even genetic strains of the same variety) can also be caused by differences in the number of copies of certain DNA sequences (in addition to point mutations changing the nucleotide sequences or crossing over of chromosomes etc.). The same phenomenon often accompanies also changes in the differentiational status of the tissue. As both genetical differences and developmental factors may lead to variations in the composition of total DNA, there is a requirement for genetically well defined biological materials and for an exactly characterized developmental status of the tissue.

Methodical Comments

First we made experiments in order to determine the extent of homology of DNA sequences in different species. As in other plant species, the reassociation kinetics of *Daucus*, *Petroselinum* and *Datura innoxia* indicated a high percentage of repeated DNA sequences amounting to about 50–60%. The calculation of genome size of one carrot variety by comparing the reassociation rate of our kinetic standard (*E. coli* DNA) with that of the unique DNA fraction indicates 5.5×10^8 nucleotide pairs per haploid genome, which corresponds to 0.6 μg of DNA per cell. This amount of DNA is quite in agreement with results based on analytical determinations (GOULD & al. 1974); the carrot genome, therefore, should contain single copy-DNA in the unique fraction (DÜHRSSEN 1979, DÜHRSSEN & al. 1979).

For a more detailed description, in particular of qualitative differences in the composition of DNA sequences of various species, carrot DNA was labelled by application of ^{3}H-thymidine to carrot tissue cultures. In order to elucidate the extent of homologous DNA sequences in carrot and *Petroselinum* or *Datura*, respectively, the ^{3}H-labelled carrot DNA was hybridized with unlabelled DNA of each of the other two species.—Voucher specimens of the plants are preserved in the tissue culture collection of our institute.

Results and Discussion

Since hybridization of ^{3}H-labelled carrot-DNA was only achieved to a limited extent of 8% in the repeated fraction and about 7-9% in the unique fraction for both species, the degree of homology seems to be rather small. Nevertheless, it has to be kept in mind that a homology of 20% of the DNA sequences of the carrot genome with the genome of *Petroselinum* still corresponds to about 10^8 nucleotide pairs per haploid genome. Whereas parsley is considered as a member of the

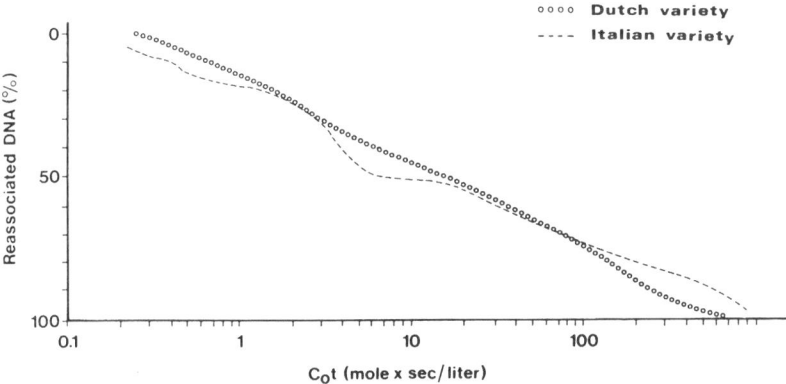

Fig. 1. Reassociation kinetics of the DNA of carrot roots of Italian and Dutch origin

Umbelliferae, i.e. the same family as carrot, *Datura innoxia* belongs to the *Solanaceae*, a family which is phylogenetically quite distant from the *Umbelliferae*. Nevertheless, it is quite remarkable that the degree of homology of *Daucus* DNA with *Petroselinum* DNA and with *Datura* DNA is more or less the same. Cot fractions of *Datura* DNA containing homologous sequences to the DNA of *Daucus* differ, however, from cot fractions of the DNA of *Petroselinum* (DÜHRSSEN & al. 1979). It cannot be decided whether these differences are due to variations in the nucleotide sequences and/or in the number of copies of homologous DNA sequences, before hybridization of labelled DNA of these particular cot fractions of *Datura* with unlabelled DNA of *Petroselinum* or *vice versa* have been made.

Small differences in reassociation kinetics in the repeated, the intermediate-repeated and the unique regions were also detected

between different varieties of *Daucus* (Schäfer & al. 1978, Schäfer 1976;
Fig. 1). In order to elucidate the homology of nucleotide sequences in the
DNA of different varieties, further hybridization experiments were
made, using ³H-labelled DNA of one variety ("Rote Riesen") with
unlabelled DNA of another variety. It was found that up to 90%
reassociation is attained in these experiments, and that most DNA
sequences of one variety can hybridize with the DNA of the other
(Duhrssen 1979). However, the number of copies varies between
varieties, and deviations occur in the Cot fractions < 0.15 and 1.5–5.0
and > 1000 (Dürssen 1979, Schäfer & al. 1978). At present no specific
functions can be assigned to this variation in the number of copies, but
future experiments have to consider that, besides the classical changes
in genome organization like chromosome mutations etc., also differences in the number of certain DNA sequences could be responsible for
the genetic variations within a species.

At present no explanation for the origin and function of such DNA-
variation in the expression of phenotypical characteristics of different
varieties can be safely offered. In the following no attempt will be
made, therefore, to associate differences in the frequency of certain
DNA sequences, which seem to be genetically fixed, with the main topic
of our investigations: the selective replication of DNA sequences during
the development of plants of the same variety.

For a general characterization of the composition of DNA of
different tissues of carrot, CsCl gradient centrifugations were carried
out. The results of these investigations indicate the occurrence of a
main band DNA in all tissues investigated with an identical density of
1.691 g/cm³ as shown in Fig. 2 and Table 1 (Schäfer & al. 1978). In the
tissues investigated, the occurence of satellites with different densities
was demonstrated ranging from a light satellite in the secondary
phloem of the root to a rather heavy satellite of 1.721 g/cm³ in carrot
callus cultures, and other satellite DNA of intermediate density. Since
the density of these satellites is an indication of their GC-content, these
results suggest changes in the composition of total DNA in tissues of
different developmental stages.

Using again an ³H-thymide labelling technique it was shown that
the preferential synthesis of a heavy satellite (1.703 g/cm³) occurs in
carrot callus cultures in log phase growth during a 24 h labelling period
(Duhrssen 1979). About 80% of the labelled DNA of this satellite
hybridized with isolated, unlabelled, highly repeated DNA with a Cot
1/2 of 0.269. Therefore, the majority of this preferentially synthesized
satellite DNA could be classified as highly repeated DNA.
Nevertheless, also main band DNA was labelled and part of this DNA
hybridized with highly repeated DNA at two Cot regions, i.e. at Cot 1/2

of 0.15 and 1.09, indicating different reassociation rates for these sequences. From these data it may be concluded that not only GC—rich sequences of repeated DNA were synthesized during this 24 h labelling period (DÜHRSSEN & NEUMANN, 1979).

Fig. 2. Analytical CsCl-density gradient profiles of DNA of secondary phloem of the root (original tissue) and cultured explants of *Daucus carota L. a* s. phloem, *b* Tissue cultures (+ kinetin)

Table 1. Buoyant density and G + C content of the DNA of various tissues of *Daucus carota L.* — M. L. = *Micrococcus lysodeicticus* DNA; PSR = pseudorabies virus DNA; sh = shoulder. — The G + C content was calculated from the buoyant density according to MANDEL & al. (1968)

DNA	Marker	density (g/cm³)			G + C content		
Root phloem	M. L.	1.691			31.6		
Leaf	M. L.	1.691	1.703		31.6	43.9	
Seedling	M. L.	1.691	sh-1.703				
					31.6	(43.9)	
Tissue culture + Kin.	PSR	1.691	1.703	1.721	31.6	43.9	62.2
Tissue culture—Kin.	PSR	1.691	1.717	1.721	31.6	58.2	62.2

Today, phytohormones are regarded as most powerful tools to influence differentiation and development of plants. The application of e.g. gibberellic acid to young carrot plants induces the transition from the vegetative rosette state to the generative state as indicated by the elongation of internodes (SCHWAB & NEUMANN 1975). Furthermore, root growth, in particular phloem-oriented cambial activity, is strongly reduced after GA₃ application. Concomitant changes in the reassociation kinetics of the DNA of the roots occur as indicated in Fig. 3 (SCHÄFER 1976, NEUMANN & al. 1977, SCHÄFER & NEUMANN 1978). Furthermore, it was shown by hybridizing one DNA fraction of unique DNA with repeated DNA that morphological changes, which were induced by GA₃, are associated with an increase of some originally unique DNA and its appearance in the intermediate-repeated DNA fraction. The results of these investigations again are an indication of a selective synthesis of certain DNA sequences in tissue of various differentiational stages (SCHÄFER & NEUMANN 1978).

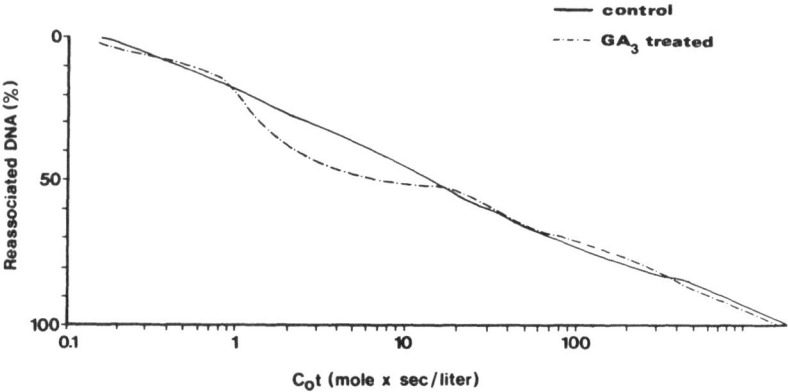

Fig. 3. The influence of GA₃ sprays on reassociation kinetics of the DNA of carrot roots

Such selective synthesis of some DNA sequences as indicated by the data presented should result in some variations in the total DNA content of cell nuclei in different tissues. Dramatic changes in the C values were e.g. found in some histological regions of *Cymbidium* protocorms at various developmental stages as originally reported by NAGL & RÜCKER (1972). The results were confirmed, in tendency, by investigations in our laboratory (unpublished). However, in most of the tissues of *Datura* or barley leaves, in which DNA determination were carried out by microfluorometric methods, the majority of cells indicated only 1 C or 2 C, and 2 C or 4 C DNA contents, in respectively haploids and diploids, corresponding to G₁ or G₂ cells, with some scattering of DNA values in between, accounting for S phase nuclei. Therefore, a correlation of a certain nuclear DNA content with differentiation seems not to be a general feature. Correspondingly, in carrot tissue cultures of various

Table 2. The influence of kinetin on turnover of DNA of carrot tissues after 2 weeks of culture. — Before labelling, the tissues were kept for 24 h in a phosphorous-free depletion medium. I = ^{32}P uptake for 6 h; II = ^{32}P uptake for 6 h followed by ^{31}P as chase for 6 h

		I c/min/mg DNA	II c/min/mg DNA
1.	— Kinetin	8767	1961
	+ Kinetin	4879	29571
2.	— Kinetin	6528	2085
	+ Kinetin	4582	20792
3.	— Kinetin	11753	3355
	+ Kinetin	6341	35617

developmental stages no clear evidence for a linear change in the DNA content was found, as should be the case after differential DNA replication. Only in rhizogenic explants some cell populations with higher C values were detected, but even these can be classified in a geometrical order indicating polyploidy (SCHÄFER & al. 1978). Thus, at least from our investigations, no unequivocal evidence for selective DNA replication is available. Two alternative interpretations can be given: Either the concentration of selectively replicated DNA is very low and beyond the sensitivity of our microfluorometric methods, or this selectively replicated DNA is readily broken down again. Reports on the occurence of metabolic DNA are available from several laboratories (SAMPSON & DAVIES 1966, ANKER & al. 1971, BRYANT & al. 1974, NEUMANN 1972, SCHÄFER & al. 1978), including our own: Table 2 shows the results of some pulse/chase experiments using ^{32}P/^{31}P, which also demonstrate the occurence of breakdown of some DNA in carrot tissue cultures.

Although some evidence for the existence of selective synthesis of certain DNA sequences is available also from other laboratories (reviewed by BUIATTI 1977), definite proof is still lacking. Moreover, neither its localization in the cell, nor its function can de defined exactly, so far. However, other studies indicate the location of different satellite DNAs in the nucleus and it is possible that they are associated with heterochromatin as postulated by INGLE & TIMMIS (1975). GUILLÉ & QUÉTIER (1973) suggested a regulatory function for such highly repeated sequences. Furthermore, some evidence exists that heavy satellites contain sequences coding for ribosomal RNA (JAWORSKI& KEY 1973, s.a. THORNBURY & SIEGEL 1973, STÄHLE & LIMA DE FARIA 1975), which may be synthesized preferentially during differentiation as has been shown by AVANZI & al. (1973).

References

ANKER, F., STROUN, M., GREPPIN, H., FREDY, M., 1971: Metabolic DNA in spinach stems in connection with ageing. — Nat. New Biol. **234**, 184—186.

BUIATTI, M., 1977: DNA amplifikation and tissue cultures. In REINERT, J., BAJAJ, Y. P. S. (Eds.): Plant cell, Tissue and Organ Culture, 358—374. — Berlin-Heidelberg-New York: Springer.

BRYANT, J. A., WILDON, D. C., WONG, D., 1974: Metabolic labile DNA in aseptically grown seedlings of *Pisum sativum L.* — Planta **118**, 17—24.

DÜHRSSEN, E., 1979: Untersuchungen zur Charakterisierung der Satelliten-DNS pflanzlicher Gewebekulturen (*Daucus carota L., Datura innoxia Mill.*). Dissertation, Universität Gießen (in preparation).

— NEUMANN, K.-H., 1979: Characterization of satellite-DNA of *Daucus carota L.* — Planta (in press).

— SAAVEDRA, E., NEUMANN, K.-H., 1979: Characterization of DNA of various plant species by reassociation kinetics and interspecies hybridization. (In press).

GOULD, A. R., BAYLISS, M. W., STREET, H. E., 1974: Studies on the growth in culture of plant cells. — J. Exp. Bot. **25**, 468—478.

GUILLÉ, E., QUETIER, F., 1973: Heterochromatic, redundant and metabolic DNA's: A new hypothesis about their structure and function. — Progr. Biophys. Mol. Biol. **27**, 123—142.

INGLE, J., TIMMIS, J. N., 1975: A role for differential replication of DNA in development. In MARKHAM, R. & al. (Eds.): Modification of the Information Content of Plant Cells, 37—52. — Amsterdam: North-Holland.

JAWORSKI, A., KEY, J. L., 1974: Distribution of ribosomal deoxyribonucleic acid in subcellular fractions of higher plants. — Pl. Physiol. **53**, 366—369.

NAGL, W., 1976: Zellkern und Zellzyklen. — Stuttgart: Ulmer.

— RÜCKER, W., 1972: Beziehungen zwischen Morphogenese und nuklearem DNS-Gehalt bei aseptischen Kulturen von *Cymbidium* nach Wuchsstoffbehandlung. — Z. Pflanzenphysiol. **67**, 120—134.

NEUMANN, K.-H., 1972: Untersuchungen über den Einfluß des Kinetins und des Eisens auf den Nucleinsäure- und Proteinstoffwechsel von Karottengewebekulturen. — Z. Pflanzenernährung Bodenkunde **131**, 211—220.

— SCHÄFER, A., DÜHRSSEN, E., 1977: Beziehungen zwischen der Zusammensetzung der DNS und der Differenzierung bei *Daucus carota*. — Chemiedozententagung Marburg, 26. März 1977.

SAMPSON, M., DAVIES, D. D., 1966: Synthesis of a metabolic labile DNA in the maturing root cells of *Vicia faba*. — Exp. Cell Res. **43**, 669—673.

SCHÄFER, A., 1976: Untersuchungen zur Charakterisierung der DNS von *Daucus carota L.* — Dissertation, Universität Gießen.

— BLASCHKE, J. R., NEUMANN, K.-H., 1978: On DNA metabolism of carrot tissue cultures. — Planta **139**, 97—101.

— NEUMANN, K.-H., 1978: The influence of gibberellic acid on reassociation kinetics of DNA of *Daucus carota L.* — Planta **143**, 1—4.

SCHWAB, B., NEUMANN, K.-H., 1975: Der Einfluß von Gibberellinsäurespritzungen auf das Blühverhalten von Karotten. — Z. Pflanzenernährung Bodenkunde **138**, 13—18.

STÄHLE, U., LIMA DE FARIA, A., 1975: Satellite DNA localization of ribosomal cistrons and heterochromatin in *Haplopappus gracilis*. — Hereditas **79**, 21—28.

THORNBURG, W., SIEGEL, A., 1973: Rapidly denatured DNAs of some higher plants. — Biochim. Biophys. Acta **312**, 211—214.

Address of the authors: ELKE DÜHRSSEN, Dr. ANGELIKA SCHÄFER, Prof. Dr. KARL HERMANN NEUMANN, Institut für Pflanzenernährung, Abteilung Gewebekultur, Justus Liebig-Universität, Eichgärtenallee 3, D-6300 Giessen, Federal Republic of Germany.

Pl. Syst. Evol., Suppl. 2, 105—110 (1979)

Department of Biology, The University of
Kaiserslautern, Federal Republic of Germany

Differential DNA Replication Involved in Transition From Juvenile to Adult Phase in *Hedera helix* (*Araliaceae*)

By

Karl-Heinz Schäffner and **Walter Nagl**, Kaiserslautern

Key Words: Angiosperms, *Araliaceae*, *Hedera.*—Morphogenesis, differential DNA replication.

Abstract: Phase change from the juvenile to the adult stage in the ivy, *Hedera helix*, is characterized by an increase in nuclear size and DNA content, and a decrease in the proportion of heterochromatin per nucleus. Our scanning cytophotometric and electron microscopic data were compared with biochemical data, and interpreted as the expression of differential replication of respectively heterochromatin and repetitive DNA during transition from the juvenile to the adult phase. The possible causal relationships between genomic diversification and phenotypic changes are briefly discussed.

The common ivy, *Hedera helix* L., shows a characteristic phase change during development. The juvenile phase is readily observed in plants growing in dark forests, or climbing on shady walls. The juvenile ivy is a creepy vine with palmate leaves growing for at least 10 years without phase change (FRANK & RENNER 1956, ROBBINS 1957), but on sunny places the plant changes to the adult form of an arborescent upright plant with entire leaves (Fig. 1), which then proceeds to flower formation and reproduction. The developmental program must be encoded in some way in the genome, but its realization is not yet understood. The phase change in the ivy comprises geotropism, leaf shape, the ability to produce adhesive roots and flowers, respectively, and metabolic events such as anthocyan synthesis. Such prominent changes must come about by a switch in the regulatory system of the genome. Actually, KESSLER & RECHES (1977) indicated that both gibberellic acid and changes in the quantity and quality of nuclear DNA are involved in phase change of the ivy. Therefore, this plant may well serve as a model object to analyze the occurrence and extent of somatic genome diversification during morphogenesis as, for instance, required by the "DNA optimization model" (Nagl 1977, 1978a, b). According to this hypothesis, the switch in the regulatory program, which evidently occurs during morphogenesis and phase change, is controlled (among other factors) by differential replication of non-coding, repetitive DNA sequences.

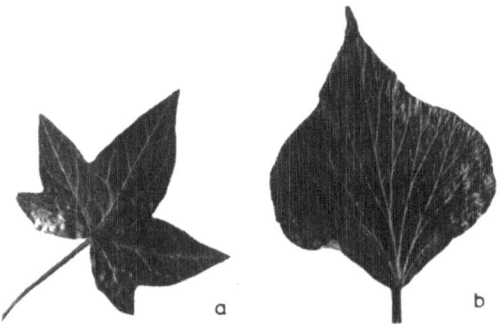

Fig. 1. Heterophylly in the ivy, *Hedera helix*. *a* Leaf of the juvenile phase. *b* Leaf of the adult phase. (0.5 ×)

Methods

Leaf buds and leaves of juvenile and adult branches of *Hedera helix* were collected from several stands and fixed with ethanol:acetic acid (3:1). After hydrolysis in 5 N HCl at room temperature for 45 min (Fox 1969), the material was Feulgen-stained, and more than 700 nuclei were individually measured with a Leitz MPV-2 scanning photometer interfaced to a DEC PDP-8/m computer. *Allium cepa* root tip telophases were used as a standard (2 C = 33.5 pg according to Van't Hof 1965).

Part of the material was fixed with glutaraldehyde and osmium tetroxide (respectively 6.25 % in Pipes buffer, pH 7.3, and 1 % in Pipes buffer, pH 6.8), and embedded according to Spurr (1969). Ultrathin sections were stained with uranyl acetate and lead citrate, and examined with a Zeiss EM 10 electron microscope.

Vouchers are preserved at the herbarium of the University of Kaiserslautern.

Results

The results of the DNA measurements are shown in Table 1 and can be summarized in the following way:

1) Juvenile buds and leaves have nuclei with a 2 C DNA content.

2) Adult buds and leaves have nuclei with a DNA content corresponding to 2 C + 71%, on average.

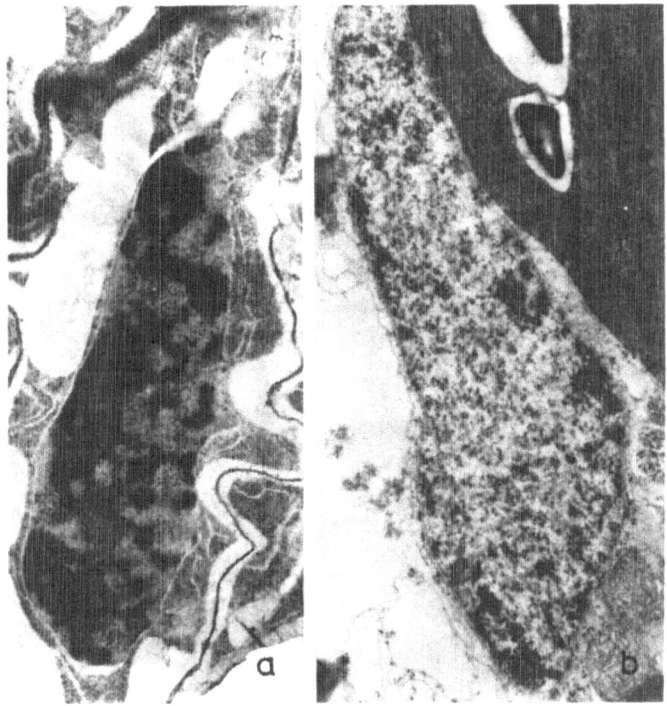

Fig. 2. Nuclear ultrastructure in (*a*) juvenile leaves, and (*b*) adult leaves of *Hedera helix*. Note that the nuclei of adult leaves are larger, but exhibit less heterochromatin than the juvenile leaf nuclei. (6,400 ×)

Table 1. The nuclear DNA content (G_1) in the bud meristem of the juvenile and adult phase of *Hedera helix* from different stands, and its interpretation in terms of differential DNA replication

Stand	Juvenile phase (pg ± S. D.)	Adult phase (pg ± S. D.)	Interpretation of adult DNA value
A	3.0 ± 0.3	5.5 ± 0.5	2 C + 82 %
B	4.0 ± 0.4	6.6 ± 0.5	2 C + 61 %
C	3.8 ± 0.3	6.4 ± 0.5	2 C + 69 %
Mean	3.6 ± 0.3	6.2 ± 0.5	2 C + 71 %

3) Statistical treatment of the data reveals that all differences recorded are significant at least at the 1 % level, some of them even at the 0.1 % level.

The increase in nuclear DNA content from the juvenile to the adult phase cannot be attributed to a shift of G_1 nuclei to the S period or G_2 phase of the cell cycle, because only the smallest values were processed, and nearly no mitotic activity occurred in the material which was collected during fall. Furthermore, the increased DNA values of adult meristems and leaf cells cannot be attributed to common endo-polyploidization, as we have never measured $4C$ but always about 30% lower values. Therefore, the higher DNA values in adult material must be the result of differential replication of the genome (Nagl 1979). In order to characterize the fractions which become replicated, other methods have to be employed. Our electron micrographs of nuclei from meristematic and mature cells of juvenile and adult leaves strongly suggest that all the variations in the nuclear DNA content are due to variation in the proportion of heterochromatin. Fig. 2 shows two extreme situations as found in meristematic nuclei of juvenile and adult leaves. Thus, we conclude that the change in nuclear DNA content between juvenile and adult branches of the ivy is due to a polyploidization step in which the heterochromatin is not participating (i.e. heterochromatin underreplication as a certain form of differential DNA replication).

Discussion

The present scanning cytophotometric and electron micros-copic study reveals that the phase change in *Hedera* is accompanied by polyploidization in combination with heterochromatin under-replication. Our data are in good agreement with those of Kessler & Reches (1977) who determined the genome size, and the number of genomes per cell, with biochemical techniques. Further characterization of the DNA fractions which are or are not replicated can be done by thermal denaturation and reassociation experiments. Such experiments were made by both Kessler & Reches (1977) and our research group (T. Müller & W. Nagl, unpubl.). The results indicate that the DNA which is extracted from adult leaves is melting at higher temperatures, i.e. it is enriched in GC-rich requences, and shows reassociation kinetics which indicate a higher proportion of repetitive DNA in comparison with DNA from juvenile leaves. Thus, GC-rich, repetitive DNA sequences located within the euchromatin are replicated, while AT-rich, less repetitive (or unique) sequences located in the heterochromatin are not replicated during phase change. Besides the fact that heterochromatin and repetitive DNA are not correlated in this species (for further examples see Nagl 1979), the question arises as whether the changes in nuclear DNA are the cause, or a consequence, of phase change in the ivy. The fact that increased DNA content and changed nuclear ultrastructure are found in buds and leaf *primordia*, i.e. before the altered morphology is visible, indicates that the switch in morphogenesis is a consequence of differential DNA replication.

Differential DNA replication such as DNA amplification and DNA underreplication evidently occur during somatic development in many plant and animal species (reviewed by NAGL 1978a; more recent results are given by ALT & al. 1978, KUENZLE & al. 1978, RENKAWITZ-POHL 1978, SCHÄFER & NEUMANN 1978, SCHÄFER & al. 1978, SCHEUERMANN 1978, STEPHEN 1978, STROM & al. 1978). In many cases non-coding repetitive DNA sequences are involved in differential replication. As repetitive DNA sequences have been assumed to play a role in the *regulation* of gene activity (BRITTEN & DAVIDSON 1969, ZUCKERKANDL 1976, and others), and as phase changes represent *regulatory problems*, it is suggestive to assume that differential DNA replication is the *cause* for changes in morphogenesis. This idea has been incorporated into the "DNA optimization model" for speciation and cytodifferentiation (NAGL 1977, 1978a). Although there is no direct proof available so far for this hypothesis, experiments should be possible to justify or falsify it. *In vitro* studies in the orchid *Cymbidium* indicated that inhibition of heterochromatin amplification prevented normal development of the protocorms (NAGL & RÜCKER 1972). Experiments are now in progress in order to analyse the importance of differential DNA replication in phase change of *Hedera helix*. The complete cytological and biochemical data will be published in a forthcoming paper.

We thank Mrs. SILVIA KÜHNER for technical assistance, and Mr. THEO MÜLLER for supply of unpublished biochemical data. A grant (Na-107/3) by the *Deutsche Forschungsgemeinschaft* is gratefully acknowledged.

References

ALT, F. W., KELLEMS, R. E., BERTINO, J. R., SCHIMKE, R. T., 1978: Selective multiplication of dihydrofolate reductase genes in methotrexate-resistant variants of cultured murine cells. — J. Biol. Chem. **253**, 1357—1370.

BRITTEN, R. J., DAVIDSON, E. H., 1969: Gene regulation for higher cells: a theory. — Science **165**, 349—357.

FOX, D. P., 1969: Some characteristics of the cold hydrolysis technique for staining tissue by the Feulgen reactions. — J. Histochem. Cytochem. **17**, 266—272.

FRANK, H., RENNER, O., 1956: Über Verjüngung bei *Hedera helix* L. — Planta **47**, 105—114.

KESSLER, B., RECHES, S., 1977: Structural and functional changes of chromosomal DNA during aging and phase change in plants. — Chrom. Today **6**, 237—246.

KUENZLE, C. C., BREGNARD, A., HÜBSCHER, U., RUCH, F., 1978: Extra DNA in forebrain cortical neurons. — Experim. Cell Res. **113**, 151—160.

NAGL, W., 1977: The DNA optimization model for speciation and cytodifferentiation. — Chrom. Today **6**, 151—152.
— 1978 a: Endopolyploidy and Polyteny in Differentiation and Evolution. — Amsterdam: North-Holland.
— 1978 b: Evidence for ontogenetic genome diversification. — Arch. Genet. **51**, 20—21.
— 1979: Condensed interphase chromatin in plant and animal cell nuclei: fundamental differences. — Pl. Syst. Evol., Suppl. **2**, 247—260.
— RÜCKER, W., 1972: Beziehungen zwischen Morphogenese und nuklearem DNS-Gehalt bei aseptischen Kulturen von *Cymbidium* nach Wuchsstoffbehandlung. — Z. Pflanzenphysiol. **67**, 120—134.
RENKAWITZ-POHL, R., 1978: Number of the repetitive euchromatic 5 S RNA genes in polyploid tissues of *Drosophila hydei*. — Chromosoma **66**, 249—258.
ROBBINS, W. J., 1957: Gibberellic acid and the reversal of adult *Hedera* to a juvenile state. — Amer. J. Bot. **44**, 743—746.
SCHÄFER, A., NEUMANN, K.-H., 1978: The influence of gibberellic acid on reassociation kinetics of DNA of *Daucus carota* L. — Planta **143**, 1—4.
— BLASCHKE, J. R., NEUMANN, K.-H., 1978: On DNA metabolism in carrot tissue cultures. — Planta **139**, 97—101.
SCHEUERMANN, W., 1978: Überprüfung der cytologischen Phänomene bei *Vicia faba*, die zu Pelc's Hypothese einer „metabolischen" DNA beitrugen. — Cytobiologie **17**, 232—245.
SPURR, A. R., 1969: A low viscosity epoxy resin embedding medium for electron microscopy. — J. Ultrastruct. Res. **26**, 31—43.
STEPHEN, J., 1978: Nucleolar DNA amplification in *Ornithogalum* endosperm. — Protoplasma **95**, 31—36.
STROM, C. M., MOSCONA, M., DORFMAN, A., 1978: Amplification of DNA sequences during chicken cartilage and neural retina differentiation. — Proc. Nat. Acad. Sci. (U.S.) **75**, 4451—4454.
VAN'T HOF, J., 1965: Relationship between mitotic cycle duration, S period duration, and the average rate of DNA synthesis in root meristems of several plants. — Experim. Cell Res. **39**, 48—58.
ZUCKERKANDL, E., 1976: Gene control in eukaryotes and the c-value paradox. "Excess" DNA as an impediment to transcription of coding sequences. — J. Mol. Evol. **9**, 73—104.

Address of the authors: KARL-HEINZ SCHÄFFNER, Prof. Dr. WALTER NAGL, Department of Biology, The University, P.O. Box 3049, D-6750 Kaiserslautern, Federal Republic of Germany.

Pl. Syst. Evol., Suppl. 2, 111—118 (1979)

Department of Biology, The University
of Kaiserslautern, Federal Republic of Germany

Extra-DNA During Floral Induction?

By

Walter Nagl, Brigitte Frisch, and **Erika Frölich,** Kaiserslautern

Key Words: Angiosperms, *Rhoeo discolor, Sambucus racemosa, Scilla decidua.* — DNA content, floral induction.

Abstract: Scanning cytophotometric DNA measurements on Feulgen-stained nuclei from vegetative and reproductive buds of various age were made in three plants, *Sambucus racemosa, Scilla decidua,* and *Rhoeo discolor.* In all three species, a certain portion of the nuclei in reproductive buds exhibited higher 2 C values than the nuclei of vegetative buds. As also telophase nuclei displayed increased DNA values, and as preliminary reassociation experiments with *Sambucus* DNA revealed remarkable differences in the cot curves of vegetative and reproductive bud DNA, the occurrence of an extra DNA (floral DNA) which is involved in the transition to flowering and the triggering of floral differentiation is discussed.

WARDELL & SKOOG (1973) reported that the DNA content of stem tissue from a flowering tobacco plant is correlated with its capacity to flower *in vitro.* Moreover, WARDELL (1976) found some floral activity of DNA extracted from flowering tobacco stems, if applied to vegetative plants. These findings indicated that flower formation can be triggered by a specific DNA fraction, which may be extra replicated (amplified) during floral induction. Therefore, we decided to study the DNA of vegetative and reproductive buds in various plants by both scanning cytophotometry of individual nuclei and denaturation and renaturation of extracted DNA. This report summarizes the cytological data, which have been obtained so far. In addition, some preliminary biochemical results are given.

Material and Methods

About 6,000 nuclei of vegetative and generative buds of three angiosperms, *Sambucus racemosa* L., *Scilla decidua,* and *Rhoeo discolor* HANCE were Feulgen-stained after cold hydrolysis with 5 N HCl for 60 min, 45 min, 30 min, and 20 min, and after hot hydrolysis with 1 N HCl for 9 min. The absolute DNA content was calculated from the extinction values according to the formula given by NAGL (1976). Measurements were made with a Leitz MPV-2 scanning cytophotometer interfaced to a Digital PDP-8 minicomputer. Various tests were made for the statistical analysis of the differences observed (for details see FRÖLICH & NAGL 1979, FRISCH & NAGL 1979). DNA of *Sambucus* was extracted according to WELLS & INGLE (1970), denatured and renatured in 0.12 M phosphate buffer, and the extinction (hyperchromicity) was recorded with a Gilford spectrophotometer type 2502, equipped with the thermoprogrammer. Cot curves were derived from the reduction in hyperchromicity according to BRITTEN & al. (1974).

Voucher specimens of the plants used are deposited in the herbarium Kaiserslautern University.

Rhoeo discolor

Independent of the type of hydrolysis used, the telophase nuclei of root tips and vegetative buds exhibited a 2 C DNA content of 16.7—16.8 pg. Telophase nuclei of floral buds showed a similar DNA content only at late stages of differentiation, while early stages displayed

Table 1. Stage- and tissue-specific 2 C values as obtained by scanning cytophotometry of telophase nuclei in *Rhoeo discolor*

Tissue and stage	DNA content (2 C, pg)	Significance of difference to vegetative buds at the 99.9 % level
Vegetative buds	16.7	
Root tips	16.8	—
Floral buds, stage 1	19.7	+
— stage 2	21.5	+
— stage 3	19.0	+
— stage 4	16.2	—
— stage 5,6	16.2	—
Flowers	16.2	—

significantly higher 2 C DNA values. The differences were significant at the 99.9% level (Table 1). Repeated experiments confirmed the observation that during early differentiation of floral buds, probably during the period of determination (transition), a transitory increase in DNA did occur in both interphase and mitotic nuclei (Figs. 1, 2). As the preferential transmission of B chromosomes was ruled out, and many theoretical considerations exclude the possibility of a changed binding capacity for the Schiff's reagent (see the discussion by FRÖLICH & NAGL 1979), the results are interpreted as the expression of some kind of extra DNA, which may be part of the floral induction system in this plant.

Scilla decidua

During a study of nuclear DNA contents and the pattern of endopolyploidy in this Liliaceous plant, a population of nuclei was detected in floral buds, flower stems, petals and carpels, which displayed

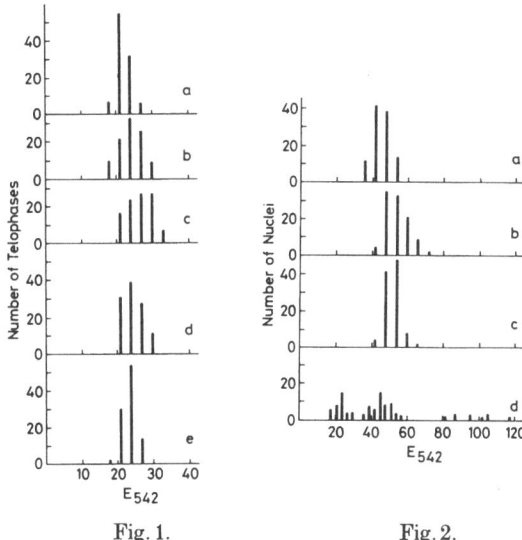

Fig. 1.   Fig. 2.

Fig. 1. Histograms of DNA contents measured in nuclei of *Rhoeo discolor* organs after hydrolysis with 5 N HCl at room temperature for 60 min. *a* Telophase nuclei of vegetative buds; *b* telophase nuclei of generative buds, stage 1; *c* telophase nuclei of generative buds, stage 2; *d*, interphase nuclei of young flowers. Note the transitory increase in DNA content

Fig. 2. Histograms of DNA contents measured in telophase nuclei of *Rhoeo discolor* after hydrolysis with 5 N HCl at room temperature for 20 min. *a* Root tip; *b* inflorescence axis of flower buds, stage 1; *c* same tissue, stage 2; *d* same tissue, stage 3; *e* same tissue, stage 4. Note the transitory increase in DNA content

a higher nuclear DNA content than nuclei of vegetative tissues (Table 2). The differences of the means were significant at the 99.95% level. Because the means were shifted to higher values also in polyploid series of DNA contents, an influence by S phase nuclei can be ruled out. As methodical errors are very unlikely (for a discussion see FRISCH & NAGL 1979), we suggest that the variation in Feulgen-extinction values actually revealed differences in the nuclear 2 C DNA content. Thus, extra DNA in floral development apparently also exists in *Scilla decidua*.

Table 2. Some examples of 2 C DNA contents in various organs of *Scilla decidua*

Organ	DNA content (2 C, pg)	Significance of difference to root tip meristem (99.95 % level)
Root tip	13.6	
Petal	16.7	+
Carpel	15.8	+
Flower stem[a]		
Population A	15.2	+
Population B	23.2	+

[a] The means of the two populations differ significantly also between each other. Endopolyploid nuclei, which are frequently found in the flower stem, display DNA values, which are multiples of either the 2 C value of population A or that of population B.

Table 3. Characterization of nuclear DNA from vegetative and floral buds of *Sambucus racemosa* (preliminary data)

Bud type	2 C value (pg)	Tm (°C)	GC (%)	Repetitive DNA (% renatured at Cot 100)
Vegetative	21.1	83.9	36	80
Floral	29.5	86.1	41	92

Sambucus racemosa

The investigation of this species is still in progress, but the preliminary results indicate that the floral buds contain more DNA than the vegetative buds in the elder, too. Scanning cytophotometric DNA measurements indicated that the average 2 C DNA content of leaf bud nuclei is 21.11 pg, while the mean in nuclei of floral buds is 29.48 pg, that is about 40 % more than in vegetative cells. As only telophases were used for the calculation, so that effects caused by S phase nuclei can be excluded.

Due to the large size of buds, the elder is well suited for biochemical DNA analyses. DNA which was extracted from vegetative and floral buds in April, differs significantly with respect to the melting temperature and thus GC content (Table 3), reassociation kinetics and thus proportion of repetitive DNA (Fig. 3), and probably also the amount of satellite DNA in CsCl density gradients (the ultracentrifugation experiments have, however, to be repeated, before exact data can be reported). Taken these findings together, clear evidence for the occurrence of extra DNA in floral buds was obtained. The reassociation kinetics, moreover, indicate that the extra DNA is repetitive and, therefore, obviously is composed of non-coding sequences.

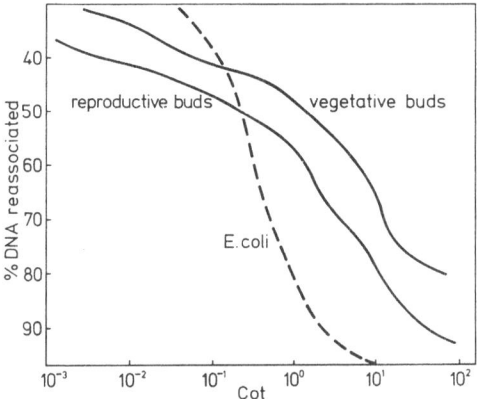

Fig. 3. Reassociation kinetics of DNA extracted from vegetative and reproductive buds of *Sambucus racemosa*; *E. coli* DNA was used as a standard. These preliminary Cot curves, which represent the means of two repeats, indicate that floral bud DNA is enriched in repetitive DNA sequences

Discussion

The data obtained strongly support the hypothesis of a "floral DNA", which is apparently extra-synthesized in the course of floral induction and floral differentiation. The hypothesis of a floral DNA, which was first put forward by WARDELL & SKOOG (1973) and WARDELL (1976), can be seen as a special case of differentiation-specific extra DNA. The amplification of certain repetitive DNA sequences was, for instance, found in chicken cartilage and retina differentiation (STROM & DORFMAN 1976, STROM & al. 1978), Guinea-pig retina (SCHMALENBERGER & NAGL 1979), *Cymbidium* protocorms (NAGL 1972, NAGL & RÜCKER 1972, 1976), and many other cases of cell differentiation and morphogenesis in plants and animals (reviewed by NAGL 1975, 1976, 1977, 1978). The accumulating evidence for differential DNA replication during differentiation falsifies the dogma of DNA constancy. According to this theory, the genome of all somatic cells is identical, and differential gene activity is established by specific regulatory proteins or RNA's, which derepress the tissue-specific genes. The appearance of control proteins according to the species-specific program is, however, not yet understood. BRITTEN & DAVIDSON (1969, 1971), ZUCKERKANDL (1976), NAGL (1978) and others put forward models, according to which the non-coding, mainly repetitive DNA sequences are involved in the control of gene activity patterns, and in the origin of phylogenetic and ontogenetic diversification (see also the introduction to this symposium given by NAGL 1979). Variation in these control sequences may actually be an important step in the regulation of speciation and cell differentiation, because it allows to escape the infinite succession of requirements of control proteins for gene activation (which themselves need the activation of other genes for their production, genes which again require control proteins to be transcribed from other activated genes, and so on). Of course, we cannot yet answer the question, how the amplification of floral (or other specific) DNA sequences is controlled, but we think that the realization of the species-specific morphology is a matter of self-instruction and self-organization of the genome, and thus independent of the genetic code per se. Experiments will have to be devised now, which may allow to confirm or to falsify this hypothesis. We do not deny the importance of environmental stimuli of floral induction, but such stimuli do not replace the interal coordination of floral differentiation.

It was shown that the transitory increase in DNA during floral differentiation can also be measured in mitotic chromosomes. This extra DNA is either integrated by an insertion/excision mechanism, or is regularly segregated during mitosis. Regulated transmission of extra DNA has, for instance, been found in the oocyte/trophocyte differentiation system of the beetle *Creophilus* (KLOC & MATUSZEWSKI 1977). Careful cytological and biochemical studies should be envisaged to discern between these possibilities.

We thank Miss BRIGITTE HOLZHAUSER for technical assistance, and the *Deutsche Forschungsgemeinschaft* for financial support (grant Na 107/3).

Note added at proof

During spring 1979, the DNA and chromatin of vegetative and floral in *Sambucus nigra* were studied in our laboratory. The results can be summarized as follows (the value for floral buds is given first, the value for vegetative buds is given in brackets):

DNA content: 30.5 pg (30.6 pg)
Tm: 84.3 °C (84.4 °C)
GC content: 36.7% (36.7%)
Repetitive DNA: at Cot 0.1: 22% (21.5%)
at Cot 1.0: 31% (28.5%)
at Cot 10.0: 43% (40.5%)

Electron-dense chromatin in electron micrographs: 22.8% (30.6%). The differences were not significant in most series of measurements. Therefore, the data obtained in *Sambucus nigra* do not confirm the hypothesis of floral DNA so far (V. BURKART, F. CSAIKL, A. DRUM, G. EBERLE, I. MICHELS, W. NAGL, N. POHLMANN, W. STEFFEN, B. STEIN, B. SCHMALENBERGER, unpublished results). However, more experiments and improved techniques are necessary in order to falsify the hypothesis of floral DNA by such negative results.

References

BRITTEN, R. J., DAVIDSON, E. H., 1969: Gene regulation for higher cells: a theory. — Science **165**, 349—357.
— — 1971: Repetitive and non-repetitive DNA sequences and a speculation on the origins of evolutionary novelty. — Quart. Rev. Biol. **46**, 111—138.
— GRAHAM, D. E., NEUFELD, B. R., 1974: Analysis of repeating DNA sequences by reassociation. — Meth. Enzymol. **29 E**, 363—406.
FRISCH, B., NAGL, W., 1979: Patterns of endopolyploid and 2 C nuclear DNA content (Feulgen) in *Scilla* (*Liliaceae*). — Pl. Syst. Evol. **131**, 261—276.
FRÖLICH, E., NAGL, W., 1979: Transitory increase in chromosomal DNA (Feulgen) during floral differentiation in *Rhoeo discolor*. — Cell Different. **8**, 11—18.
KLOC, M., MATUSZEWSKI, B., 1977: Extrachromosomal DNA and the origin of oocytes in the telotrophic-meroistic ovary of *Creophilus maxillosus* (L.) (*Staphyliniidae, Coleoptera/Polyphaga*). — W. Roux's Arch. **183**, 351—368.
NAGL, W., 1972: Evidence of DNA amplification in the orchid *Cymbidium in vitro*. — Cytobios **5**, 145—154.
— 1975: Organization and replication of the eukaryotic chromosome. — Progr. Botany **37**, 186—210.
— 1976: Zellkern und Zellzyklen. — Stuttgart: Ulmer.
— 1977: Organization and replication of the eukaryotic chromosome. — Progr. Bot. **39**, 132—152.
— 1978: Endopolyploidy and Polyteny in Differentiation and Evolution. — Amsterdam: North-Holland.
— 1979: Search for the molecular basis of diversification in phylogenesis and ontogenesis. — Pl. Syst. Evol., Suppl. **2**, 3—25.
— RUCKER, W., 1972: Beziehungen zwischen Morphogenese und nuklearem DNS-Gehalt in aseptischen Kulturen von *Cymbidium* nach Wuchsstoffbehandlung. — Z. Pflanzenphysiol. **67**, 120—134.
— — 1976: Effects of phytohormones on thermal denaturation profiles of *Cymbidium* DNA: Indication of differential DNA replication. — Nucl. Acids Res. **3**, 2033—2039.

SCHMALENBERGER, B., NAGL, W., 1979: Different DNA content, chromatin condensation, and transcription activity in retina cell nuclei of the Guinea-pig. — Pl. Syst. Evol., Suppl. **2**, 119—125.

STROM, C. M., DORFMAN, A., 1976: Amplification of moderately repetitive DNA sequences during chick cartilage differentiation. — Proc. Nat. Acd. Sci. (U.S.) **73**, 3428—3432.

— MOSCONA, M., DORFMAN, A., 1978: Amplification of DNA sequences during chicken cartilage and neural retina differentiation. — Proc. Nat. Acad. Sci. (U.S.) **75**, 4451—4454.

WARDELL, W. L., 1976: Floral activity in solutions of DNA extracted from tobacco stems. — Pl. Physiol. **57**, 855—861.

— 1977: Floral induction of vegetative plants supplied with a purified fraction of DNA from stems of flowering plants. — Pl. Physiol. **60**, 885—891.

— SKOOG, F., 1973: Flower formation in excised tobacco stem segments. III. DNA content in stem tissue of vegetative and flowering tobacco plants. — Pl. Physiol. **52**, 215—220.

WELLS, R., INGLE, J., 1970: The constancy of the buoyant density of chloroplast and mitochondrial DNAs in a range of higher plants. — Pl. Physiol. **46**, 178—179.

ZUCKERKANDL, E., 1976: Gene control in eukaryotes and the C-value paradox. "Excess" DNA as an impedient to transcription of coding sequences. — J. Mol. Evol. **9**, 73—104.

Address of the authors: Prof. Dr. WALTER NAGL, BRIGITTE FRISCH, and ERIKA FRÖLICH, Department of Biology, The University, P.O. Box 3049, D-6750 Kaiserslautern, Federal Republic of Germany.

Pl. Syst. Evol., Suppl. 2, 119—125 (1979)

Department of Biology, The University
of Kaiserslautern, Federal Republic of Germany

Different DNA Content, Chromatin Condensation, and Transcription Activity in Retina Cell Nuclei of the Guinea-Pig *

By

Bernd Schmalenberger and **Walter Nagl**, Kaiserslautern

Key Words: Guinea-pig. — DNA Content, nuclear structure, retina.

Abstract: Nuclei of the cornea and of several cell types occurring in the retina (pigmented epithelium, bipolar cells, rods and cones, and Müller cells) were studied with respect to their DNA content. chromatin ultrastructure, and RNA synthesizing activity. Significant differences were found. It is concluded that transcription activity depends on the proportion of decondensed chromatin, and that both may be controlled by differential DNA replication during early development of the eye.

The retina as the innermost layer of the eye may well deal as a model for studies of cell differentiation and functional control in mammals, because several different cell types develop close together, such as the cells of the pigmented epithelium, the rods and cones, bipolar cells, ganglion cells, and the Müller (glial) cells. The individual cell types are well described and distinguishable in both light and electron micrographs (e.g. MAGALHAES & COIMBRA 1972, RASMUSSEN 1973, BÜLOW 1975. KANEKO & al. 1976, NGUYEN-LEGROS 1978).

Methods

In this study the following techniques were used in order to find relationships between nuclear DNA content, chromatin organization and RNA synthesis: scanning cytophotometry of Feulgen-stained nuclei, ^{3}H-thymidine and ^{3}H-uridine autoradiography, light and electron microscopy of nuclei after

---

* A preliminary report.

Table 1. Characterization of nuclear organization and activity in several cell types of the Guinea-pig retina and cornea (as standard)

Code	Cell type	DNA content (2C, pg) ± S.D.[a]	Significance of differences[b] (99.9% level)	Per cent chromatin condensed[c]	³H-uridine incorporation (number of silver grains)[d] ± S.D.
C	Cornea cells	5.9 ± 0.3	C to all others	ca. 2	not studied
P	Photoreceptor cells	6.2 ± 0.3	P:B, P:M, P:E, P:C	71.1 ± 4.4	30.1 ± 8.7
B	Bipolar cells	6.3 ± 0.3	B:P, B:M, B:E, B:C	14.9 ± 4.8	59.8 ± 11.6
M	Müller cells	4.9 ± 0.5	M:B, M:C, M:E	ca. 1[e]	?[f]
E	Pigment epithelium	6.5 ± 0.4	E:M, E:C, E:B, E:P	ca. 34	?[f]

[a] Number of nuclei measured: 50–100 of each type. [b] At least at the 99% level (most differences are significant at the 99.9% level). [c] Calculated from electron micrographs after weighing of 10 nuclei of each group, and separately of the cut-out condensed chromatin. If fewer nuclei were analyzed, only "ca.-values" are given. [d] Mean of 100 nuclei. [e] The euchromatin of Müller cell nuclei is, however, more densely arranged than that in other cell types. [f] In these cell types there is a large-scale variation of silver grains, which is not yet understood.

acetic ethanol and glutaraldehyde/OsO$_4$ fixation, respectively. For DNA measurements, the tissue was hydrolyized with 5 N HCl at room temperature for 45 min and Feulgen stained, using chicken erythrocytes as an external standard (2 C = 2.8 pg) and Guinea-pig cornea cells as an internal standard (2 C = 5.9 pg). Incubation in 20 μCi/ml ^{3}H-thymidine and 20 μCi/ml ^{3}H-uridine into excised eyes was made in BME medium supplemented with 6 % fetal calf serum and 200 μg/ml penicillin/streptomycin for 90 min. Exposure time was 10 days to 4 months.

Characterization of the Nuclei

The results are summarized in Table 1. The nuclear DNA content is significantly different between the cell types, except between the cells of the pigmented epithelium and the bipolar cells. As these differences may be explained by DNA amplification during the differentiation period of the eye, ^{3}H-thymidine experiments were performed to detect the stage of DNA extra synthesis. The retina nuclei of adult animals did not incorporate the radioactive precursor. Also the retina nuclei of 61-64 days old intra-uterine embryos did not synthesize DNA *in vitro*. Many of the nuclei of 15-17 days old embryos, whose eyes were incubated *in vitro*, were, however, labelled. As also mitotic figures were detected, the labelled nuclei may be in a mitotic S-period. Some of the larger nuclei were particularly heavily labelled, what could be the expression of an endo-S-period. Unfortunately, loci of DNA amplification could not identified in this stage, so that further experiments are necessary to elucidate the evolution of the differencies in nuclear DNA content.

Electron micrographs of ultrathin sections revealed significant differences in the proportion of condensed chromatin (Table 1). Müller cell nuclei show a homogeneous distribution of euchromatin and exhibit some small areas with condensed chromatin, scattered throughout the nucleus. The nuclei of photoreceptor cells (rods and cones) normally display two large patches of condensed chromatin, while nuclei of the bipolar cells show a characteristic central location of condensed chromatin (called "fried-egg nuclei" in our lab; Fig. 1). The differences in the amount of condensed chromatin (varying between 14% and 71 %) are much larger than the differences in the nuclear DNA content, so that most of this condensed chromatin must be inactivated euchromatin (for a discussion of this aspect see NAGL 1979).

The number of silver grains after ^{3}H-uridine incorporation was evaluated in a semiquantitative way in each 100 nuclei of the photoreceptor cells and the bipolar cells (Table 1; incorporation of the precursor into other cell types showed to high variation for calculation of a suggestive mean). The differences found are due to the differences in the proportion of condensed chromatin, as only decondensed chromatin

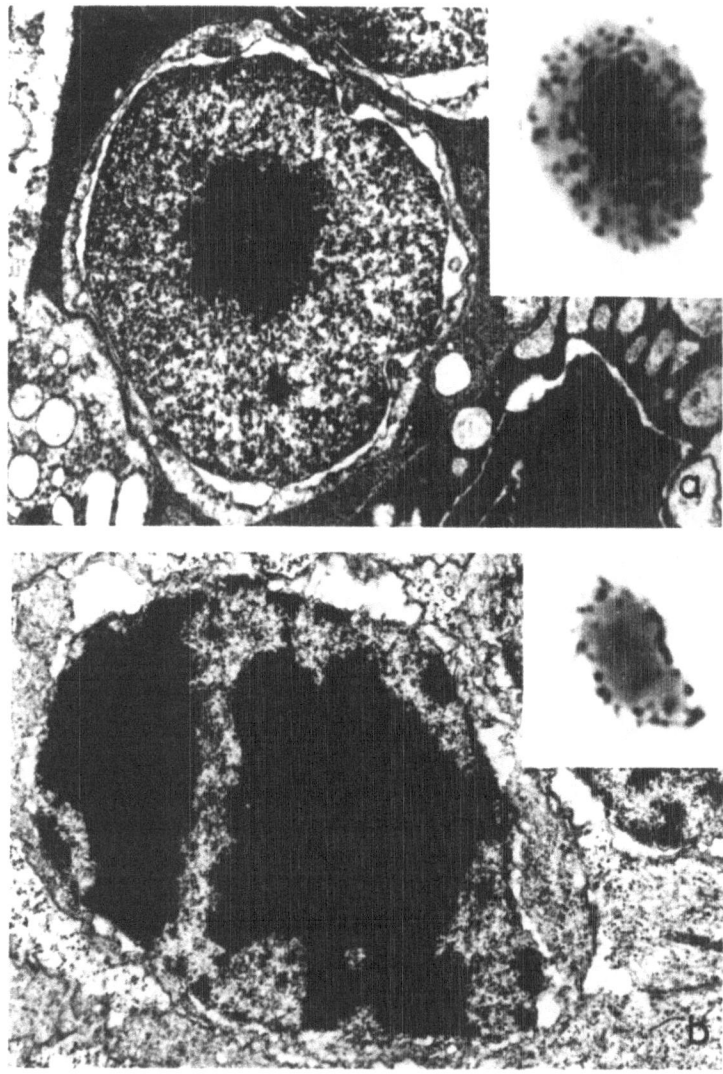

Fig. 1. Electron micrographs and light microscope autoradiograms of retina
cell nuclei of the Guinea-pig. *a* Nucleus of a bipolar cell (6,720 ×); insert: ^{3}H-
uridine incorporation into decondensed chromatin. *b* Nucleus of a photoreceptor cell (9,800 ×); insert: ^{3}H-uridine incorporation into the peripheral
uncondensed chromatin layer only

was labelled. This indicates that differential chromatin condensation actually represents a mechanism to control transcription activity in certain cells.

Discussion and Conclusions

The main finding of this investigation is that the nuclei of various cell types in the Guinea-pig retina differ in their DNA content, the degree of chromatin condensation, and their transcription activity. How are these parameters related to each other? A negative correlation between the degree of chromatin condensation and the rate of RNA synthesis is well established (HSU 1962, ALLFREY & al. 1963, NAGL 1969, AUER & ZETTERBERG 1972, see also the reviews by NAGL 1976, 1979). It is also well established that functional chromatin condensation occurs through changes in the composition and modification of chromosomal proteins (histones and non-histones; PAUL & GILMOUR 1968, SMART & BONNER 1971, JOHNS 1972, MIRSKY & SILVERMAN 1973, McCARTHY & al. 1974, THOMAS & SCHRAM 1977, reviews: BUSCH 1978, Cold Spring Harbor Symp. Quant. Biol. 1978). However, the control of such changes in chromosomal proteins, and thus in chromatin condensation and decondensation, is only poorly understood. Modifying enzymes, and gene-specific non-histone chromosomal proteins have to be transcribed themselves. This transcription need specific chromatin decondensation for itself, which has to be controlled by other non-histones or protein-modifying enzymes. These again must be transcribed from decondensed chromatin and so on. Thus succession can only find an end, if there is a control mechanism, which is independent from that protein regulatory system as described. Attempts to find such an independent control mechanism have been made by BRITTEN & DAVIDSON (1969, 1971), who suggested that the non-coding repetitive DNA sequences play some role therein (see also ZUCKERKANDL 1976). Various types of changes in the nuclear DNA content and composition as a self-organizing and self-instructing control mechanism of cell differentiation and morphogenesis have been envisaged by NAGL (1977, 1978) and summarized in the "DNA optimization model". According to this, amplification and other kinds of differential replication of non-coding control DNA could be a basic step in cytodifferentiation. Actually, somatic DNA amplification and differential replication of non-coding DNA sequences have been recently found (reviewed by NAGL 1978). The differences in the nuclear DNA content of certain retina cell types as described in this paper support the hypothesis that differential DNA replication may play a regulatory role in retina cell differentiation. This interpretation is consistent with findings in the chicken, were repetitive DNA sequences are amplified during cartilage and retina differentiation (STROM & al. 1978), and with extra DNA in forebrain cortical neurons (Kuenzle & al. 1978). Further examples are given in this volume (Nagl & al. 1979, Schäffner & Nagl 1979).

We speculate, therefore, that the specific transcription activity of certain retina cell types is due to specific degrees of chromatin condensation. The latter is controlled via the chromosomal protein metabolism by cell differentiation prior to cell function. Cell differentiation, however, is a consequence of the self-instructing and self-organizing realiziation of the *"Bauplan"*, which is encoded by control DNA, via differential DNA replication. The retina may well deal as a model system to prove or falsify this working hypothesis.

We thank the *Deutsche Forschungsgemeinschaft* for support (grant Na-107/3), and Mr. K.-H. Schäffner for scanning the nuclei for DNA determinations.

References

Allfrey, V. G., Littau, V. C., Mirsky, A. E., 1963: On the role of histones in regulating RNA synthesis in the cell nucleus. — Proc. Nat. Acad. Sci. (U.S.) **49**, 414—421.

Auer, G., Zetterberg, A., 1972: The role of nuclear proteins in RNA synthesis. — Exp. Cell Res. **75**, 245—253.

Britten, R. J., Davidson, R. H., 1969: Gene regulation for higher cells: a theory. — Science **165**, 349—357.

— — 1971: Repetitive and non-repetitive DNA sequences and a speculation on the origins of evolution novelty. — Quart. Rev. Biol. **46**, 111—138.

Bulow, N., 1975: The retinal pigment epithelium and photoreceptor cells on monkey eyes. — Cell Tiss. Res. **161**, 521—540.

Busch, H., (Ed.), 1978: The Cell Nucleus, vol. 4 & 5: Chromatin. — New York: Academic Press.

Cold Spring Harbor Symposia Quant. Biol. **42**: Chromatin (1978).

Hsu, T. C., 1962: Differential rate in RNA synthesis between euchromatin and heterochromatin. — Exp. Cell Res. **27**, 332—334.

Johns, E. W., 1971: Histones, chromatin structure and RNA synthesis. — Nature New Biol. **237**, 87—88.

Kaneko, A., Lam, D. M. K., Wiesel, T. N., 1976: Isolated horizontal cells of elasmobranch retinae. — Brain Res. **105**, 567—572.

Kuenzle, C. C., Bregnard, A., Hübscher, U., Ruch, F., 1978: Extra DNA in forebrain cortical neurons. — Exp. Cell Res. **113**, 151—160.

Magalhaes, M. M., Coimbra, A., 1972: The rabbit retina Müller cell. A fine structural and cytochemical study. — J. Ultrastruct. Res. **39**, 310—326.

McCarthy, B. J., Nishiura, J. T., Doenecke, D., Nasser, D. S., Johnson, C. B., 1974: Transcription and chromatin structure. — Cold Spring Harbor Symp. Quant. Biol. **38**, 763—771.

Mirsky, A. E., Silverman, B., 1973: Effect of selective extraction of histones on template activities of human chromatin by use of exogenous DNA and RNA polymerases. — Proc. Nat. Acad. Sci. (U.S.) **70**, 1973—1975.

NAGL, W., 1969: Correlation of structure and RNA synthesis in the nucleolus-organizing polytene chromosomes of *Phaseolus vulgaris*. — Chromosoma **28**, 85—91.

— 1976: Nuclear organization. — Ann. Rev. Pl. Physiol. **27**, 39—69.

— 1977: The DNA optimization model for speciation and cytodifferentiation. — Chrom. Today **6**, 151—152.

— 1978: Endopolyploidy and Polyteny in Differentiation and Evolution. — Amsterdam: North-Holland.

— 1979: Condensed interphase chromatin in plant and animal cell nuclei: fundamental differences. — Pl. Syst. Evol., Suppl. **2**, 247—260.

— FRISCH, B., FRÖLICH, E., 1979: Extra DNA synthesis involved in floral differentiation. — Pl. Syst. Evol., Suppl. **2**, 211—218.

NGUYEN-LEGROS, J., 1978: Fine structure of the pigment epithelium in the vertebrate retina. — Intern. Rev. Cytol., Suppl. **7**, 464—499.

RASMUSSEN, K. E., 1973: A morphometric study of the Müller cells, their nuclei and mitochondria in the rat retina. — J. Ultrastruct. Res. **44**, 96—112.

PAUL, J., GILMOUR, R. S., 1968: Organ-specific restiction of transcription in mammalian chromatin. — J. Molec. Biol. **34**, 305—316.

SCHÄFFNER, K.-H., NAGL, W., 1979: Differential DNA replication involved in transition from juvenile to adult phase in *Hedera helix*. — Pl. Syst. Evol., Suppl. **2**, 105—110.

SMART, J. E., BONNER, J., 1971: Study on the role of histones in relation to the template activity and precipitability of chromatin at physiological ionic strengths. — J. Molec. Biol. **58**, 675—684.

STROM, C. M., MOSCONA, M., DORFMAN, A., 1978: Amplification of DNA sequences during cartilage and neural retina differentiation. — Proc. Nat. Acad. Sci. (U.S.) **75**, 4451—4454.

THOMAS, C., SCHRAM, A., 1977: Correlation between the condensation state of rDNA and rRNA synthesis during *Xenopus laevis* oogenesis. — Biol. Cell **30**, 49—54.

ZUCKERKANDL, E., 1976: Gene control in eukaryotes and the C-value paradox. "Excess" DNA as an impedient to transcription of coding sequences. — J. Mol. Evol. **9**, 73—104.

Address of the authors: BERND SCHMALENBERGER, Prof. Dr. WALTER NAGL, Department of Biology, The University, P.O. Box 3049, D-6750 Kaiserslautern, Federal Republic of Germany.

Gene Numbers and Transcription

Pl. Syst. Evol., Suppl. 2, 129—140 (1979)

Institute of Botany, The University of
Hannover, Federal Republic of Germany

Gene Number Estimates in Plant Tissues and Cells

By

Manuel Kiper, Dorothea Bartels, and
Heinrich Köchel, Hannover

Key Words: Angiosperms, *Petroselinum crispum, Hordeum vulgare.* — DNA
reassociation, DNA/RNA hybridization, mRNA complexity.

Abstract: Two methods are described to measure numbers of active genes.
The one uses ^{3}H-unique DNA saturation and the other the kinetics of
hybridization of ^{3}H-cDNA with its complementary RNA to estimate mRNA
complexities (informational lengths) and hence gene numbers. We used both
approaches to measure numbers of active genes in plants. The data obtained
were 11,500-13,300 in parsley root callus and 9,200-10,000 in parsley leaf,
14,200 m barley young leaves. In barley and in parsley a close correspondence
between complexity of poly(A)$^-$mRNA and poly(A)$^+$mRNA was found.

The haploid nuclear DNA content of flowering plants varies from less
than 1 pg to more than 100 pg (NAGL 1978) the coding capacity of which
exceeds the information necessary for more than 10^6 different proteins.
Genetic considerations suggest that in higher organisms a maximum of
up to 50,000 structural genes may be expressed during life time(OHTA &
KIMURA 1971), a value now approved for higher animals by some direct
measurements. As higher plants undergo sophisticated differentiation
processes it is interesting to see which amount of genetic information
they use for establishing the function of differentiated cells and tissues.
To investigate this we used DNA/RNA-hybridization techniques both
with ^{3}H-unique DNA and ^{3}H-cDNA and polysomal mRNA excess
(GALAU & al. 1974, BISHOP & al. 1974).

The rationale of the unique DNA saturation measurement is
displayed in Fig. 1. Sequence complexities, i.e. the informational length
of polysomal mRNA can be determined by measuring the percentage of

unique nuclear DNA of known informational length which can be rendered double-stranded when annealed to total polysomal RNA in RNA excess. The unique DNA complexity is determined by measuring the haploid genome length in nucleotides and multiplying with the

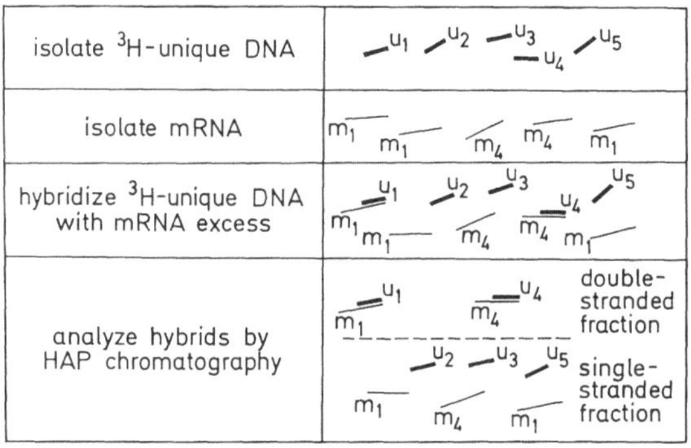

Fig. 1. Hybridization of ^{3}H-unique DNA in mRNA excess

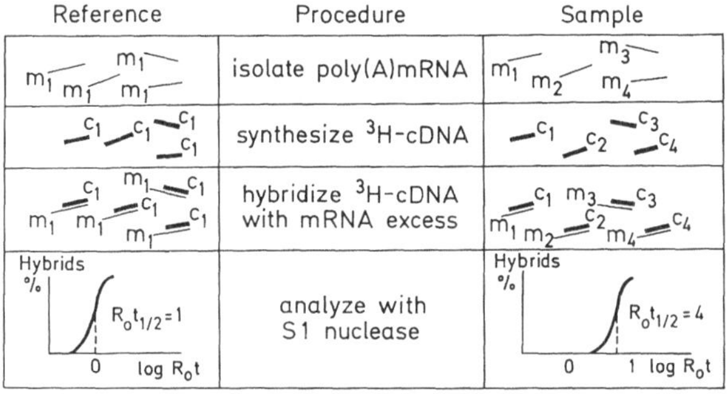

Fig. 2. Hybridization of ^{3}H-cDNA in poly(A)mRNA excess

fraction of unique DNA in the genome. Gene numbers then will result by dividing the mRNA complexity in nucleotides by the average length of mRNA.

The rationale behind the cDNA method is to compare the kinetics of hybridization of an unknown population mRNA/cDNA with a standard mRNA/cDNA the complexity of which is known. As is shown in Fig. 2 it will take four times as long to reach the point of half hybridization ($R_0t_{1/2}$)

if the complexity of the mRNA population is increased respectively. Thus the $R_0t_{1/2}$ of the reaction compared with a known standard will give the complexity of the mRNA population and by dividing by the average length of all mRNA provide an estimate of the number of genes.

It is clear that both methods will only estimate the number of genes represented in polysomal RNA, i.e. structural genes. Furthermore, both methods are confined to individual cells and tissues and will not give an estimate of the maximum number of genes ever active during differentiation. Values given must thus be taken as minimum estimates.

For **Materials and Methods** see the accompanying paper of BARTELS & KIPER.

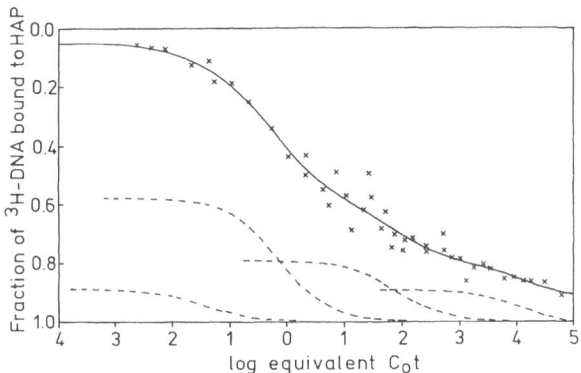

Fig. 3. Renaturation kinetics of total *Hordeum vulgare* DNA (300 nucleotides). — The curve through the crosses is a computer least square fit for a maximum of for second order components given as subcurves. Table 1 lists the kinetic parameters of the components

Results

Renaturation Kinetics of Total Nuclear DNA and Preparation of ^{3}H-Unique DNA. Barley and parsley total DNA were prepared as described in methods. Fig. 3 represents the renaturation kinetics of 300 N long nuclear barley DNA, its kinetic parameters are listed in Table 1. It is seen that only a small percentage of the whole genome is single-copy DNA the other being repetitive. The reassociation kinetics of 300 N long parsley nuclear DNA is quantitatively described elsewhere (KIPER & HERZFELD 1978). The dashed line in Fig. 5 represents the ideal renaturation kinetics. Unique DNA was prepared as described in methods. The reassociation of the single-copy ^{3}H-DNA in the presence of an excess of unfractionated 700 N long total nuclear DNA is shown in Fig. 4 for barley and Fig. 5 for parsley. By C_0t 50,000 78.9 % of the

Table 1. Analysis of the barley genome

Component	Fraction of fragment (%)	Chemical complexity (nucleotide pairs per haploid genome)	Kinetic complexity (nucleotide pairs per haploid genome)	Repetition frequency (per haploid genome)	Rate constant $K \, (M \times sec/1)$
Foldback	5.0	2.9×10^8			
Highly repetitive	11.2	6.6×10^8	2.3×10^3	293,000	29.21
Fast intermediate repetitive	42.3	2.5×10^9	1.8×10^5	14,000	1.412
Slow intermediate repetitive	20.9	1.2×10^9	1.1×10^7	138	0.0138
Unique	11.9	7.0×10^8	7.0×10^8	1	0.0001

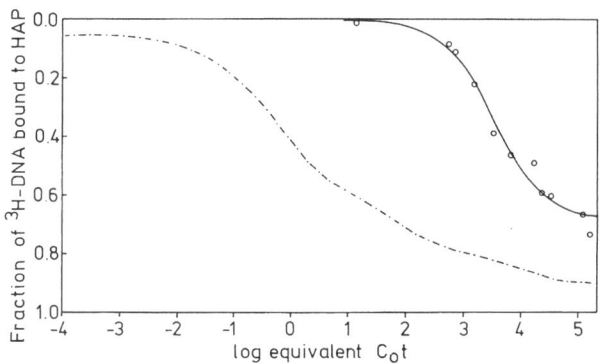

Fig. 4. Reassociation of barley unique DNA. The o represent the reassociation kinetics of 700 N total nuclear DNA with trace amounts of ^{3}H-unique DNA (250 300 N long). The curve is a least squares fit to the data with a plateau value of 68.3%. The rate constant K was $0.296 \times 10^{-3} M^{-1} sec^{-1}$. Included is the ideal reassociation kinetics of total 300 N long nuclear DNA (dashed line)

parsley unique DNA had reacted and 68.3% of the barley unique DNA at the end of the reaction. Figs. 4 and 5 indicate that the single-copy ^{3}H-DNAs used in this study contain no detectable contamination with repetitive sequences. In the absence of driver DNA less than 0.4% of the ^{3}H-DNA binds to hydroxyapatite at C_0t below 1; up to 0.7% binds at the highest C_0ts measured with excess tRNA (50,000–100,000 Msec/1). These values were routinely substracted from the hybridization data obtained with excess RNA. We take the unique DNA to represent a sequence complexity of 12 % that of the chemical complexity (genome size) of the whole haploid genome. Though the parsley genome contains 30 % unique DNA (KIPER & HERZFELD 1978) and the barley genome 25 % (KÖCHEL 1978) most of these sequences are rather short. As derived in

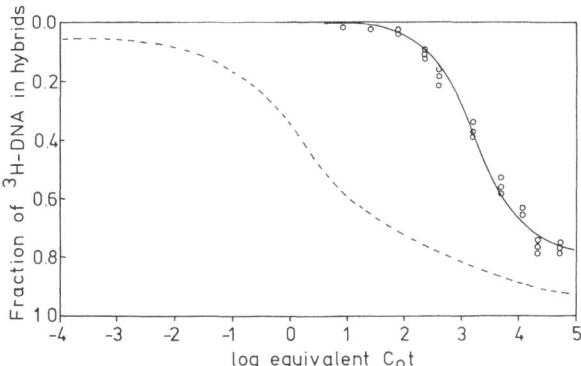

Fig. 5. Reassociation of parsley unique DNA. The o represent the reassociation kinetics of 700 N total nuclear DNA with trace amounts of ^{3}H-unique DNA (250-300 N long). The curve is a least squares fit to the data with an RMS of 0.1 % and a plateau value of reactability of 78.9 %. The rate constant K was $0.56 \times 10^{-3} M^{-1} sec^{-1}$. Included is the reassociation kinetics of total 300 N long nuclear DNA (dashed line) quantitatively described elsewhere (KIPER & HERZFELD 1978).

length elsewhere (KIPER 1979) we take the ^{3}H-unique DNA both in parsley and in barley to be a representative probe of the long unique sequences revealed in reassociation kinetics of 300 N long DNA, i.e. 12 %

Sequence Complexities of Polysomal mRNA as Measured by ^{3}H-Unique DNA Saturation. Sequence complexities of total polysomal mRNA and of polysomal poly(A)mRNA were determined by measuring the percentage of the unique ^{3}H-DNA described above which could be rendered doublestranded when annealed to total polysomal RNA in RNA excess (GALAU & al. 1974). As displayed in Fig. 6 for barley 1 % of the unique ^{3}H-DNA had hybridized with excess RNA at saturation. As can be seen by the identical saturation values for poly(A)mRNA and total polysomal mRNA poly(A)[-]mRNA comprises no additional sequence complexity to that contained in poly(A)[+]mRNA. As is

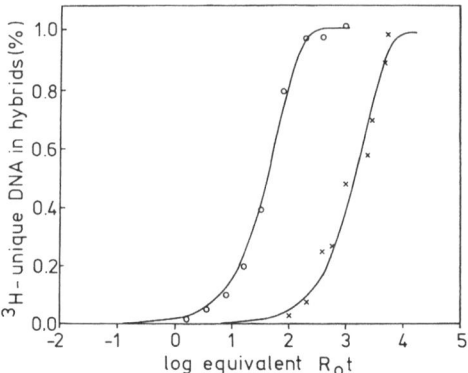

Fig. 6. Hybridization of barley unique DNA with polysomal mRNA. The curves represent the hybridization of unique ^{3}H-DNA with an excess of polysomal poly(A)mRNA (o) and an excess of total polysomal RNA($\times$). Saturation values were 1.01 and 1.00%, respectively, rate constants were $1.63 \times 10^{-2} M^{-1}sec^{-1}$ and $0.475 \times 10^{-3} M^{-1}sec^{-1}$, respectively

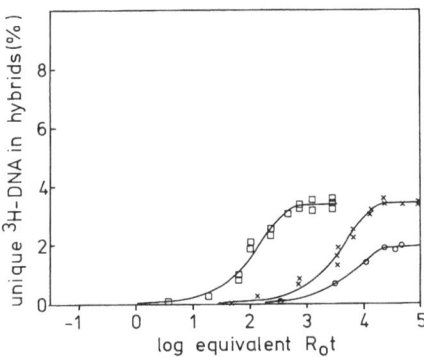

Fig. 7. Hybridization of parsley unique ^{3}H-DNA with polysomal mRNA. — The curves represent the hybridization of unique ^{3}H-DNA with an excess of polysomal poly(A)mRNA($\square$) and an excess of total polysomal RNA ($\times$) of root callus and of total leaf polysomal RNA (o). Saturation values were as indicated in Fig. 2

displayed in Fig. 7 this also holds for parsley root callus mRNA, where about 3.4% of our unique DNA preparation was driven in hybrids by excess RNA at saturation. For comparison a measurement of parsley leaf mRNA sequence complexity is included in Fig. 7. It can be seen that saturation is reached at a lower level than with root callus cells. The data

Table 2. Sequence complexities of polysomal mRNAs as measured by hybridization to ^{3}H-unique DNA

RNA		Saturation value (%)	Corrected saturation value (%)[a]	Complexity[b] (nucleotides)	Number of sequences[c]
Parsley	root callus polysomal poly(A)mRNA	3.35	4.25	1.9×10^7	13,300
Parsley	root callus polysomal mRNA	3.43	4.35	2.0×10^7	13,700
Parsley	leaf polysomal mRNA	2.45	3.11	1.4×10^7	10,000
Barley	young leaves polysomal poly(A)mRNA	1.01	1.46	2.2×10^7	14,400
Barley	young leaves polysomal mRNA	1.00	1.46	2.1×10^7	14,300

[a] Corrected for a 78.9% reactability of the parsley unique DNA-tracer and a 68.3% reactability of the barley unique DNA-tracer.
[b] Referring to a unique DNA sequence complexity of 2.3×10^8 Np/haploid parsley genome and a unique DNA sequence complexity of 7.3×10^8/haploid genome.
[c] Assuming asymmetric transcription and a number average length of 1,400 N for parsley mRNA and 1,500 N for barley mRNA.

are compiled in Table 2 together with gene numbers derived from these results assuming asymmetric transcription and 1400 M mRNA median length.

Sequence Complexities and Abundancies of Polysomal Poly(A)mRNA as Measured by Hybridization With Complementary ^{3}H-cDNA. To ascertain the gene numbers reported above by an independent method and to provide some insight into the range of mRNA abundance classes in parsley root callus cells and leaf, poly(A)mRNA was hybridized to complementary ^{3}H-cDNA, and the extent of hybridization was monitored by resistance to S 1 nuclease (BISHOP&al. 1974). The rationale of this method was outlined in Fig. 2. The results are presented in Fig. 8 and Fig. 9. The dashed line represents the hybridization of rabbit globin mRNA with its labelled cDNA as a kinetic standard with which to

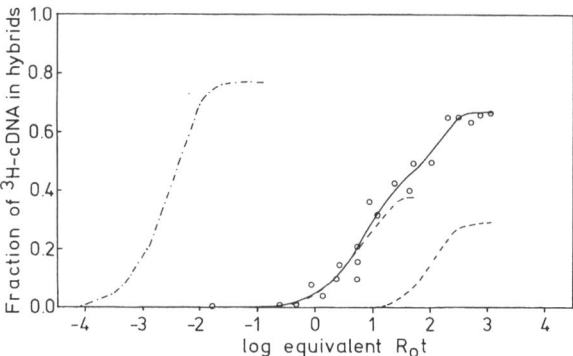

Fig. 8. Hybridization between parsley root callus polysomal poly(A)mRNA with its complementary ^{3}H-cDNA. — The reactions were analyzed using S 1 nuclease, and the line through the data represents a computer least square fit to the data using two components (dashed lines). The use of a third component did not improve the solution. The dashed-pointed line represents the kinetics of hybridization of globin mRNA with its complementary cDNA. The parameters are listed in Table 3

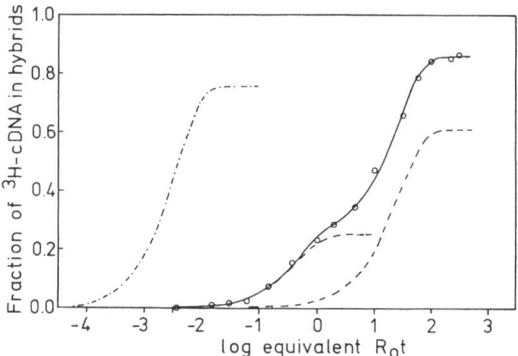

Fig. 9. Hybridization between parsley leaf polysomal poly(A)mRNA with its complementary cDNA. — See legend of Fig. 8

compare more complex hybridization systems. Rabbit globin mRNA which consists of α and β globin sequences with a total complexity of 4×10^5 daltons (WILLIAMSON & al. 1971) hybridizes with a rate constant K of $260 \, M^{-1} sec^{-1}$ under our experimental conditions. It is clear from Fig. 8 that the reaction of parsley poly(A)mRNA with its homologous cDNA is not a simple pseudo-first-order kinetic. Computational analysis showed that the data are best described by two components representing

Table 3. Abundance classes in parsley polysomal poly(A)mRNA

Poly(A)mRNA source	Fraction of polysomal poly(A)RNA mass[a]	K (M^{-1}sec^{-1})	K$_{pure}$[b] (M^{-1}sec^{-1})	Number of 1,400 N sequences[c]	Number of molecules per cell per sequence[d]
Root callus	0.56	0.1328	0.235	960	60
	0.44	0.0094	0.022	10,500	4
Leaf	0.29	2.34	8.07	28	1,060
	0.71	0.020	0.028	9,200	8

[a] The fraction of polysomal poly(A)RNA mass is calculated as the individual component to the total reacting ^{3}H-cDNA.

[b] K$_{pure}$ is calculated diving K with the fraction of polysomal poly(A)RNA.

[c] The number of molecules represented by each transition was calculated referring to the complexity of globin mRNA (WILLIAMSON & al. 1971) and to the rate constant of hybridization of rabbit globin ^{3}H-cDNA (600 N) with its complementary poly(A)mRNA found to be 260 M^{-1}sec^{-1} under our hybridization conditions. As the parsley cDNA preparations were of equal length we directly compared its K$_{pures}$ with that of globin ^{3}H-cDNA/poly(A)mRNA hybridization.

[d] In parsley root cells we determined the ratio of poly(A)mRNA to DNA to be 0.08 pg/4.0 pg = 0.02. This amounts to about 100,000 mRNA sequences of length 1,400 N per cell. Copies per sequence per cell then was calculated as

$$\frac{\text{fraction of polysomal poly(A)RNA} \times 10^5}{\text{number of 1,400 N sequences}}.$$

different abundance classes. It should be stressed that the data could be resolved into a larger number of components but this would not significantly improve the degree of fit. Furthermore, the division into abundance classes is not quantitatively exact in the sense, that within a given class, all sequences will have the same abundance exactly. As is listed in Table 3 the greatest contribution to the total base sequence complexity from the RNA population investigated is from the least abundant resolvable class which is the most slowly hybridizing component. Comparing the K of both components with that of globin we conclude that there are two different abundant classes in parsley root callus suspension cells one with 56% of all mRNAs representing about 960 different sequences of average length 1,400 N and the other comprising 44% and representing about 10,500 different species. Frequencies were 60 per cell for the more abundant and 4 per cell for the less abundant mRNA class. As displayed in Fig. 9 and listed in Table 3 in leaf we found two abundance classes one comprising 29% and the other 71% of all mRNAs representing about 30 and 9,200 different species each being present 1,060 and 8 times per cell, respectively, on an average.

Discussion

Polysomal Poly(A)⁻mRNA Contains No Sequences Different From Polysomal Poly(A)⁺mRNA.

Polysomal Poly(A)$^-$mRNA Contains No Sequences Different From Polysomal Poly(A)$^+$mRNA. As is displayed in Figs. 6 and 7 both in barley and in parsley sequence complexities of poly(A)$^+$mRNA and of total mRNA are not differing significantly indicating that poly(A)$^-$ mRNA contains no sequences different from polysomal poly(A)$^+$mRNA. As only 20–40% of all polysomal mRNA is poly(A)$^+$mRNA (Ragg & al. 1977) and as the rate constants of hybridization of both RNA populations differ by about a factor 30 we conclude that our poly(A)$^+$mRNA preparations are purified about 100 fold compared to total polysomal mRNA preparations. As the hybridization reactions are driven by the least abundant and most complex class of mRNA any large contamination of poly(A)$^+$mRNA by a complex poly(A)$^-$mRNA population must reflect in a retarded fraction of the hybridization kinetics. As we have no evidence for this we conclude that poly(A)$^-$mRNA is not a class of mRNA contributing a separate sequence complexity to the overall polysomal sequence complexity. This accords with the finding of Ragg & al. (1975) that in parsley root callus suspension cells both the poly(A)$^+$ and the poly(A)$^-$ polysomal mRNA translated in a cell free reticulocyte translation system stimulated the synthesis of proteins with similar size distribution; an individual protein, phenylalanine ammonia lyase, was detected in both polysomal RNA fractions.

mRNA Abundancies in Root Callus and Leaf Reflect Two Different States of Differentiation.

mRNA Abundancies in Root Callus and Leaf Reflect Two Different States of Differentiation. Taken together the results described above imply that gene numbers in root callus and in leaf differ by about 30% (11,500/9,200 = 1.25; 13,700/10,000 = 1.37) as measured by both methods. However, there is a major difference in what concerns abundancies of individual mRNAs. Whereas in root callus cells a broad spectrum of about 960 mRNA species make up the most abundant mRNA class, in leaf there exists a rather narrow class of about 30 different species making up 30% of all mRNAs and each being present more than 1,000 times per cell as contrasted to the other 70% being present only 8 times per cell on an average. These differences between root callus cells and leaf cells may really reflect two different states of differentiation as it can be correlated with different spectra of proteins, too. Whereas in leaf cells as visualized by Page a minor number of prominent protein bands can be distinguished (Apel & Kloppstech 1978) in root callus no prominent bands are formed but rather a continuum of many different proteins (Schröder, personal communication). Since root callus cells are fast proliferating we assume that these numerous abundant proteins are confined to growth and division processes, whereas in leaves the few most abundant mRNAs are involved in manufacturing proteins for establishing photosynthetic active structures (APEL & KLOPPSTECH 1978).

Plant Tissues Contain About 10,000 to 15,000 Active Structural Genes. As compiled in Table 2 and Table 3 in different plants and different tissues with two different methods we were able to reveal similar mRNA sequence complexities and gene numbers. Though drastic differences may appear in the number of individual sequences per cell, the total number of different active genes does not vary to a great extent. In all tissues we revealed the presence of 10,000 to 15,000 different 1,400 N sized polysomal mRNAs. Though this only is a minimum estimate of the total number of genes of higher plants, it may well be that major differentiation processes occur by changes in the relative amount of individual sequences and will only marginally effect an overall change of the whole mRNA population as was demonstrated for the sea urchin (GALAU & al.1976). It hence may turn out that gene number estimates of tissues as given above represent the majority of all structural genes present in higher plants. Recently, GOLDBERG & al.(1978) determined the gene number in tobacco leaf. Their estimate was 12,000 with the cDNA method and 27,000 with unique DNA method. As discussed elsewhere (KIPER 1979) we believe that the high gene number of 27,000 is an overestimate due to an invalid assumption reduced to 15,000 under the assumption underlying our ^{3}H-unique DNA preparations and comparing favorably with the 12,000 genes found in tobacco leaves with the cDNA method (GOLDBERG & al. 1978).

Gene numbers estimated for plants here lie well within gene numbers determined for higher animals (LEWIN 1975). They prove that higher plants, too, have at their disposal in individual cells and tissues a mass of information for structural protein diversification equaling that of intricuate differentiated animal tissues.

This work was supported by a grant to Dr. HERZFELD by the DFG.

References

APEL, K., KLOPPSTECH, K., 1978: The plastid membranes of barley (*Hordeum vulgare*). — Eur. J. Biochem. **85**, 581—588.

BISHOP, J. O., MORTON, J. G., ROSBASH, M., RICHARDSON, M., 1974: Three abundance classes in HeLa cell messenger RNA. — Nature **250**, 199—204.

GALAU, G. A., BRITTEN, R. J., DAVIDSON, E. H., 1974: A measurement of the sequence complexity of polysomal messenger RNA in sea urchin embryos. — Cell **2**, 9—20.

— KLEIN, W. H., DAVIS, M. M., WOLD, B. J., BRITTEN, R. J., DAVIDSON, E. H., 1976: Structural gene sets active in embryos and adult tissues of the sea urchin. — Cell **7**, 487—505.

GOLDBERG, R. B., HOSCHEK, G., KAMALAY, J. C., 1978: Sequence complexity of nuclear and polysomal RNA in leaves of the tobacco plant. — Cell **14**, 123—131.

KIPER, M., 1979: Gene number as measured by single-copy DNA saturation with mRNA are routinely overestimates. — Nature **278**, 279—280.

KIPER, M., HERZFELD, F., 1978: DNA sequence organization in the genome of *Petroselinum sativum (Umbelliferae)*. — Chromosoma (Berl.) **65**, 335—351.

KÖCHEL, H., 1978: Sequenzkomplexität der mRNA von *Hordeum vulgare* L. (Gerste). — Diplomarbeit, TU Hannover.

LEWIN, B., 1975: Units of transcription and translation: the relationship between heterogeneous nuclear RNA and messenger RNA. — Cell **4**, 11—20.

NAGL, W., 1978: Endopolyploidy and Polyteny in Differentiation and Evolution. — Amsterdam: North-Holland.

OHTA, T., KIMURA, M., 1971: Functional organization of genetic material as a product of molecular evolution. — Nature **233**, 118—119.

RAGG, H., SCHRÖDER, J., HAHLBROCK, K., 1975: Poly(A)-containing RNA from *Petroselinum hortense*: isolation, properties and messenger function *in vitro*. — Mol. Biol. Reports **2**, 119—127.

— — — 1977: Translation of poly(A)-containing and poly(A)-free messenger RNA for phenylalanine ammonia-lyase, a plant specific protein, in a reticulocyte lysate. — Biochim. Biophys. Acta **474**, 226—233.

WILLIAMSON, R., MORRISON, M., LANYON, G., EASON, R., PAUL, J., 1971: Properties of mouse globin messenger ribonucleic acid and its preparation in milligram quantities. — Biochemistry **10**, 3014—3021.

Address of the authors: Dr. MANUEL KIPER, DOROTHEA BARTELS, and HEINRICH KÖCHEL, Institut für Botanik der Universität, Herrenhäuserstr. 2, D-3000 Hannover, Federal Republic of Germany.

Pl. Syst. Evol., Suppl. 2, 141—149 (1979)

Institute of Botany, The University of
Hannover, Federal Republic of Germany

DNA Sequence Representation in RNA of the Higher Plant *Petroselinum crispum*

By

Dorothea Bartels and Manuel Kiper, Hannover

Key Words: *Petroselinum crispum.* — Repetitive DNA, single-copy DNA, transcription, rRNA, poly (A)mRNA.

Abstract: The genome of parsley (*Petroselinum crispum*) is made up of different sequence classes; i.e. various repetitive classes and unique DNA. By DNA/RNA hybridization techniques we investigated which DNA classes code for the different RNA species. Ribosomal RNA is transcribed from middle repetitive sequences. On the basis of the hybridization data we calculated 2,880 ribosomal genes per haploid genome. Poly(A)mRNAs are only coded for by unique DNA sequences. Preliminary results show that poly(A) hn RNA differing from poly(A)mRNA in size, labelling kinetics, and sequence complexity, is coded for by unique sequences, too.

Materials and Methods

Growth of Cells and Labelling of RNA *in vivo*. Freely suspended callus cells originating from root explants of *Petroselinum crispum* (MILL.) A. W. HILL were cultured under sterile conditions in a synthetic medium as described by SEITZ & RICHTER (1970).

To achieve high specific activities of the labelled polysomal RNAs cells were given 250 µCi/ml ^{3}H-uridine.

Preparation of Total Nuclear DNA. DNA from *Petroselinum* was prepared as previously described (KIPER & HERZFELD 1978); for hybridization assays the DNA was further purified by incubation for 20 h at 37 °C in 0.4 M NaOH.

Preparation of [3]-H-cDNA. ^{3}H-cDNA was synthesized from polysomal poly(A)RNA according to the procedure of FRIEDMAN & ROSBASH (1977). The reaction mixture (100 µl) consisted of 0.05 M Tris-HCl pH 8.3, 0.06 M MgCl$_2$, 0.02 M dithiothreitol, 100 µg/ml Actinomycin D, 5 µg/ml oligo(dT)$_{12-18}$. unlabelled deoxynucleotides (dATP, dGTP, dTTP) 0.001 M, 25 µCi (25.5 Ci/M) ^{3}H-dCTP, 4 µg poly(A)-RNA, 10 U reverse transcriptase. After incubation at 37 °C for 1 h, 5 µl of 10 % SDS was added and the reaction kept for 5 min at 37 °C.

The cDNA was treated with 0.4 N NaOH for 5 min at 100 °C, neutralized and applied to a Sephadex G-50 column equilibrated with 0.12 M Phosphatebuffer, 0.5 % SDS. The fractions containing TCA precipitable radioactivity were pooled.

Preparation of Polysomal Poly(A)mRNA and Ribosomal RNA. Polysomes were prepared as described by PFISTERER & KLOPPSTECH (1977). RNA was isolated on a cesium chloride gradient as described for nuclear RNA from nuclei. Total polysomal RNA was fractioned by oligo-dT-cellulose chromatography into ribosomal RNA and poly(A)mRNA, as described above.

Preparation of Poly(A)nuclear RNA (hnRNA) and Oligo-dT-chromatography. Nuclei from parsley were prepared according to a modified procedure from TAUTVYDAS (1971). Cells were washed with buffer A consisting of 0.05 M Tris-HCl pH 7.5, 0.02 M $MgSO_4$ and then suspended in 4 % gum arabic solution (GAS) (GAS solution: 4 % GAS, 0.15 M sucrose, 4 mM Mg-acetate, 5 mM 2-mercaptoethanol 5 mM Mes buffer pH 6.1). Pure gum arabic solution was purified before use by centrifugation at 14,000 × g for 1 hr, then by filtration through nylonscreens and sterilised. The cell suspension was homogenized in a 60 ml potter (Braun, Melsungen). The homogenate was filtered through a series of nylonscreens up to pore diameters of 10 µm. The filtrate was centrifuged at 1,000 × g for 10 min to pellet the nuclei. For further purification the nuclei were centrifuged through a gum arabic gradient ranging from 12 % GAS to 4 % GAS.

RNA was obtained to a modified procedure of GLISIN & al. (1974) by lysing the nuclear fraction in RM medium: 0.1 M Tris-HCl pH 8.0, 1 % (w/v) sarcosyl, 0.01 M EDTA, 0.1 M NaCl, 10 µg/ml Proteinase K. Up to here all manipulations took place at 0 °C. Then 1 g/ml CsCl was added and mixed. The solution was layered onto a cushion of 2.4 ml 5.7 M CsCl with 0.1 EDTA (density 1,707) in a cellulosenitrate tube. 0.5 ml of RM medium was then layered on the top. The gradient was centrifuged in a Beckman SW 40 rotor at 28,000 rpm for 14 h at 20 °C. The supernatant with the banded polysaccharides, DNA and proteins was removed, and the RNA pellet resuspended in 0.5 ml NETSP buffer (0.5 M NaCl, 10 mM EDTA, 10 mM Tris-HCL pH 7.5, 0.5 % sodium dodecylsulfate, 10 µg/ml Proteinase K). The RNA was incubated for 3 min at 65 °C and applied to an oligo-dT-cellulose column equilbrated with NETSP. After extensively washing with the same buffer the poly(A)RNA was eluted with ETSP (10 mM EDTA, 10 mM Tris-HCL pH 7.5, 0.5 % sodium dodecylsulfate, 10 µg/ml Proteinase K). The pooled RNA was precipitated by the addition of absolute ethanol. After 10-20 h at — 20 °C the precipitate was pelleted, dried and suspended in a suitable buffer.

In vitro **Labelling of hnRNA.** To obtain hnRNA with high specific radioactivity *in vivo* labelled hnRNA was labelled with di-[3] H-methyl sulfate, additionally. The *in vitro* labelling was carried out in a similar way as described by HEIKKILA & al. (1977). The poly(A)RNA sample was dissolved in 0.1 M sodium phosphate buffer (pH 7.8) and then added to the original ampoule of di-^{3}H-methyl sulfate (2.5 mCi; 3.95 Ci/mmol). After incubation for 2 h at 4 °C, the reaction was terminated by the addition of 1 ml 0.1 M sodium phosphate buffer (pH 7.8). The unreacted di-^{3}H-methyl sulfate was removed by chromatography on Sephadex G-50. The labelled RNA was precipitated with 2 volumes of 95 % ethanol.

Assay of Hybridization and Reassociation. Annealing of DNA was performed as described previously (KIPER & HERZFELD 1978). DNA/RNA hybridization reactions were carried out in 0.12 M phosphate buffer/0.5 % SDS at 60 °C or 0.4 M phosphate buffer/0.5% SDS at 64 °C in sealed capillaries. The resulting C_0t or R_0t values were corrected to 0.18 M Na^+ as suggested by BRITTEN & al. (1974). RNA and DNA concentrations were measured spectrophotomatrically assuming 1 OD = 40 µg/ml RNA and 50 µg/ml DNA. Maximum DNA and RNA concentrations used were 8 mg/ml. Tracer reactions were driven by an at least 1,000 fold excess of the other component.

DNA reassociation and single-copy DNA/RNA excess hybridization reactions were analyzed by HAP chromatography as described by KIPER (1978) with the modification that the phosphate buffer was made 0.5 % SDS.

cDNA/RNA excess hybridization reactions were analyzed by S 1 nuclease assay. Buffer conditions were those of DAVIDSON & BRITTEN (1973). 25 µg/ml native and 25 µg/ml denatured calf thymus DNA were added, the sample divided into two equal aliquots to one of which was added sufficient S 1 nuclease to digest a 20 fold excess of DNA to acid solubility. Both aliquots were incubated for 45 min at 37 °C. The reactions were terminated by chilling to 4 °C; both samples were then analyzed for cold TCA precipitable radioactivity. Percent duplex was calculated as cpm (sample with S 1) to the cpm (sample without S 1). Tracer zero time values (2–8%) were substracted from the data. RNA/DNA excess hybridization reactions were analyzed by diluting the samples into a large volume 2 × SSC buffer. The probes were divided into two aliquots. One aliquot was exposed to 20 µg/ml pancreatic RNase and 5 units T_1 RNase. Then both aliquots were incubated at 37 °C for 1 h. Yeast RNA was added as a carrier (10 µg/ml), and the samples were adjusted to 10 % trichloroacetic acid. The acid precipitable radioactivity in the RNase treated aliquot relative to that in the aliquot not treated was used to calculate the percent of the RNA hybridized. Control experiments were carried out to correct the intrinsic RNase resistance of the RNA preparations, which ranged from 2–6%.

Computer Analysis. Second order reactions (DNA reassociation, DNA excess hybridization) and first order reactions (RNA excess reactions) were analyzed by a least square computer program.

Abbreviations Used: DNA = deoxyribonucleic acid; EDTA = ethylene diamidetetraacetate; N = nucleotides; RNase = ribonuclease; SDS = sodium dodecylsulfate; SSC = 0.15 M sodium chloride plus 0.015 M sodium citrate; TRIS = 2-amino-2-(hydroxymethyl)-1,3propandiol.

In *Petroselinum crispum* studies by DNA/DNA reassociation kinetics revealed 70 % of DNA to be repetitive sequences differing in repetion frequency from about 40 to more than 10^5 per haploid genome and 30% to be unique sequences (KIPER & HERZFELD 1978), the majority of which consists of sequences shorter than 500 N (KIPER, submitted).

Little information exists regarding the transcriptional processes in plants. Higher plant genomes are similar in size, complexity, and sequence organization to those of animals (KIPER & HERZFELD 1978, WALBOT & DURE 1976, ZIMMERMANN & GOLDBERG 1977), but a special feature of plant genomes is their high proportion (60 %) of repeated DNA sequences (FLAVELL & al. 1974). The high content of repetitive sequences raises the question whether there exists a substantial class of structural genes being repetitious in the genome besides the ribosomal RNAs. To answer this question hybridization of an excess of DNA with RNAs of high specific radioactivity (MELLI & al. 1971) and with labelled cDNAs prepared from their corresponding RNAs can be employed (BISHOP & al. 1974, ROSBACH & al. 1974). The proportion of RNA in DNA/RNA hybrids was determined by low salt incubation with RNase and that of cDNA in cDNA/DNA hybrids by S 1 nuclease.

For several reasons plant cell cultures of parsley, *Petroselinum sativum*, growing heterotrophically in a liquid medium (SEITZ & RICHTER 1970) were chosen for these experiments. Large quantities of metabolically active cells can be easily obtained, under well-defined and sterile conditions. DNA and RNA can be labelled *in vivo* to high specific activity by supplying radioactive precursors to the liquid growth medium. Furthermore there occurs no large interference with chloroplasts nucleic acids.

Results and Discussion

Ribosomal RNA. Fig. 1 shows the hybridization kinetics of sheared highly labelled ribosomal RNA with total nuclear DNA (700 N long). It reveals that ribosomal RNA hybridizes with middle repetitive sequences: Rate constant $K = 0.074$ M^{-1} sec^{-1} corresponding log equivalent C_0t $1/2$ of 1.1. From these data the number of ribosomal genes can be calculated. The rate constant of hybridization compared to unique DNA directly measures genome frequencies of hybridized components (MELLI & al. 1971). Unique DNA hybridizes with total DNA (700 N) with a kinetic rate constant of 1.8×10^{-4} M^{-1} sec^{-1}. The rate constant of the repetitive ribosomal RNA and the single-copy DNA were compared: $0.074/0.00018 = 411,1$. Ribisomal RNA is for the factor $411,1$ more frequent compared with unique DNA, which is found once in the genome. Because DNA/RNA hybridization kinetics are retarded compared with DNA/DNA reassociation the number must be corrected by multiplying with a factor of 3.5 (CHAMBERLAIN & al. 1978). So we calculate a number of $2,880$ genes per telophase nucleus. This result is in fair agreement with other findings, showing that in a lot of plants ribisomal genes range from $1,250$ to $31,900$ per haploid genome (INGLE & al. 1975).

Messenger RNA. Since parsley cell cultures offer the great advantage of getting nucleic acids of high specific activity by *in vivo* labelling (up to 10^5 cpm/µg RNA) two different experimental approaches could be applied to examine the DNA sequence representation in poly(A)mRNA.

At first trace amounts of ^3H-polysomal poly(A)mRNA, several times chromatographed on oligo dT cellulose, were hybridized to

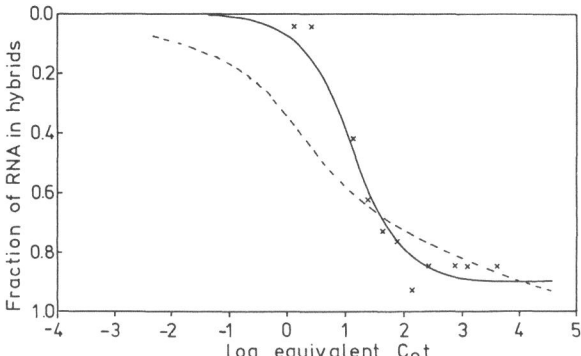

Fig. 1. Hybridization of ribosomal RNA with excess DNA (——— × ———) shows the hybridization kinetics of sheared ribosomal RNA with total nuclear 700 N long DNA. The samples were assayed with RNase as outlined in "Materials and Methods". The percentage of labeled RNA in hybrid was corrected for background RNase resistance. The dashed line represents the DNA/DNA reassociation kinetics of total nuclear DNA used in these experiments

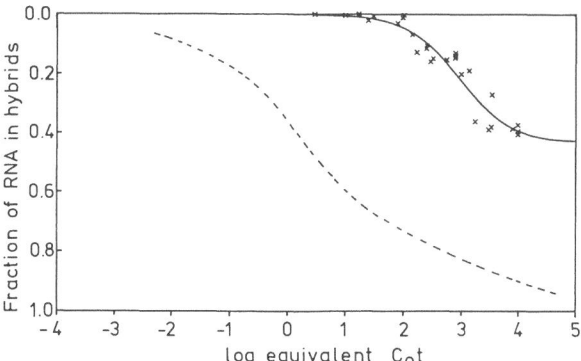

Fig. 2. Hybridization of ^{3}H-poly(A)mRNA with excess DNA Trace amounts of polysomal poly(A)mRNA were hybridized to an excess of nuclear DNA as described in Materials and Methods. The RNA/DNA hybrids were analyzed by treatment with RNase A and T_1, followed by TCA precipitation. The curve (——— × ———) through the data points represents the best least square solution of the data with one component. A two component solution does not improve the fit. For dashed line see Fig. 1

an excess of sheared nuclear DNA (700 N). The results of this DNA/RNA hybridization experiment are shown in Fig. 2. All the RNA hybridized with kinetics ($K = 0.9 \times 10^{-3}\,M^{-1}\,sec^{-1}$; $C_0t\,1/2 = 1.1 \times 10^3$) indicating that it was complementary to, and hence transcribed from single-copy DNA. The failure to obtain complete hybridization of the poly(A)mRNA is most probably an effect of insufficient DNA excess for the most abundant mRNAs (GOLDBERG & al. 1973) and of mRNA instability at the elevated temperatures employed.

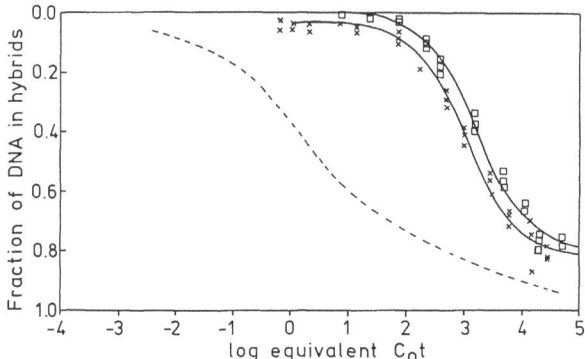

Fig. 3. Reassociation of cDNA with nuclear DNA ³H-cDNA synthesized from poly(A)mRNA was annealed in a vast excess of sheared total nuclear DNA. Hybridization was performed as described in experimental procedures. The curve (——— × ———) through the data points represents the least square fit for one component. A two component solution does not improve the fit. (——— ———) represents the reassociation kinetics of ³H-single-c opy DNA (300 N) with an excess of total nuclear DNA (700 N). For dashed line see Fig. 1

In another assay the kinetics of reassociation of cDNA from poly(A)mRNA with an excess of nuclear DNA was determined. The results are presented in Fig. 3. The DNA reassociated as a single second order reaction with a K of $0.9 \times 10^{-3}\,Mol^{-1}\,sec^{-1}$ and a $C_0t\,1/2$ of 1.1×10^3 M × sec/l. The reassociation kinetics of ³H-single-copy DNA (300 N long) with an excess of 700 N total nuclear DNA is also shown ($K = 0.56 \times 10^{-3}\,Mol^{-1}\,sec^{-1}$). Since the ratio of fragment lengths of cDNA to single-copy DNA is about two the result of hybridization kinetics clearly indicates that all of the polysomal poly(A)mRNA is transcribed from single-copy DNA.

Concerning the sequence representation of poly(A)mRNA similar experiments with tobacco plants (GOLDBERG & al. 1978) showed the same results as obtained with parsley cell cultures. Investigations with animal and fungi, however, showed that besides the single-copy DNA also a small repetitive component was coding for poly(A)mRNA (CAMPO & BISHOP 1974, FIRTEL & al. 1972; HEIKKILA & BROWN 1977, KLEIN & al. 1974, TIMBERLAKE & SHUMARD 1977). As mentioned above more than 60% of the plant genome is made up of repetitive sequences. Thus it is very surprising that all mRNA seems to be coded for by single-copy DNA sequences. Investigations concerning the sequence complexity of

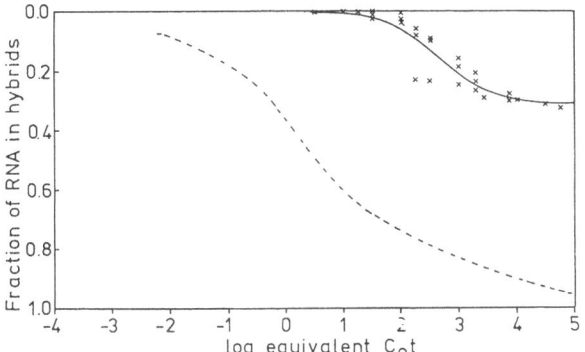

Fig. 4. Hybridization of ³H-poly(A)hnRNA with nuclear DNA. Hybridization of *in vitro* labelled hnRNA with excess nuclear DNA was performed as described in experimental procedures. The solid line through the data points represents the least square fit for one component. A two component solution does not improve the fit. For dashed line see Fig. 1

polysomal mRNA proved that the poly(A) fraction of mRNA is a representative probe of all mRNAs (see the paper of KIPER & al. in this volume).

Heterogeneous Nuclear RNA. We were successful in isolating heterogeneous poly(A) nuclear RNA (hnRNA) defined by its fast kinetics of labelling, its size distribution other than mRNA, its sequence complexity greatly exceeding that of mRNA (KIPER & al., in press). Investigations with animal cells showed that besides unique sequences repetitive sequences are represented in hnRNA of animals, too. Because of the difficulty to isolate hnRNA from plant cells, up to now nothing is known about sequence representation in plant cell tissues. To get high specific activity *in vivo* labelled hnRNA from parsley cultures was given an additional *in vitro* label. As is demonstrated in Fig. 4 hybridization experiments with a excess of DNA proved that the mass of hnRNA is coded for by single-copy DNA sequences, too.

Our experiments indicate that except of the ribosomal RNAs all RNA species investigated, i.e. mRNA and hnRNA are coded for by single-copy DNA. Though this makes up only a fraction of all the unique DNA (see the accompanying paper of Kiper & al. in this volume) the function of the mass of the repetitive DNA sequences in the genome is unclear at all and waits for further experiments.

References

Bishop, J. O., Morton, J. G., Rosbash, M., Richardson, M., 1974: Three abundance classes in HeLa cell messenger RNA. — Nature **250**, 199—204.

Britten, R. J., Graham, D. E., Neufeld, B. R., 1974: Analysis of repeating DNA sequences by reassociation. — In Grossmann, L., Moldave, K., (Eds.): Methods in Enzymology, Nucleic Acids and Protein Synthesis Vol. **29**, Part E, pp. 363—406.

Champo, M. S., Bishop, J. O., 1974: Two classes of messenger RNA cultured rat cells: repetitive sequence transcripts and unique sequence transcript. — J. Molec. Biol. **90**, 649—663.

Chamberlain, M. E., Galau, G. A., Britten, R. J., Davidson, E. H., 1978: Studies on nucleic acid reassociation kinetics: V. Effects of disparity in tracer and driver fragment lengths. — Nucl. Acids Res. **5**, 2073—2094.

Davidson, E. H., Britten, R. J., 1973: Organization, transcription, and regulation in the animal genome. — Quart. Rev. Biol. **48**, 555—613.

Flavell, R. B., Bennett, M. D., Smith, J. B., Smith, D. B., 1974: Genome size and the proportion of repeated nucleotide sequence DNA in plants. — Biochem. Genet. **12**, 257—269.

Firtel, R. A., Jacobsen, A., Lodish, H. F., 1972: Isolation and hybridization kinetics of messenger RNA from *Dictyostelium discoideum*. — Nature New Biol. **239**, 225—228.

Friedman, E. Y., Rosbash, M., 1977: The synthesis of high yields of full-length reverse transcripts of globin mRNA. — Nucl. Acids Res. **4**, 3455—3471.

Glisin, V., Crkvenjakov, R., Byus, C., 1974: Ribonucleic Acid isolation by cesium chloride centrifugation. — Biochem. **13**, 2633—2637.

Goldberg, R. B., Galau, G. A., Britten, R. J., Davidson, E. H., 1973: Nonrepetitive DNA sequence representation in sea urchin embryo messenger RNA. — Proc. Nat. Acad. Sci. (U.S.) **70**, 3516—3520.

— Hoschek, G., Kamalay, J. C., 1978: Sequence complexity of nuclear and polysomal RNA in leaves of the tobacco plant. — Cell **14**, 132—131.

Heikkila, J. J., Brown, J. R., 1977: Analysis of rabbit brain polysomal poly(A)+mRNA by DNA excess hybridization. — Biochim. Biophys. Acta **474**, 141—153.

Ingle, J., Timmis, J. N., Sinclair, J., 1975: The relationship between satellite deoxyribonucleic acid, ribosomal ribonucleic acid gene redundancy, and genome size in plants. — Pl. Physiol. **55**, 496—501.

Kiper, M., 1979: Gene numbers as measured by single-copy DNA saturation with mRNA are routinely overestimates. — Nature **278**, 279—280.

KIPER, M., 1978: A quick hydroxyapatite chromatography technique especially adapted for work with DNA networks. — Analyt. Biochem. **91**, in the press.

— HERZFELD, F., 1978: DNA sequence organization in the genome of *Petrolselinum sativum* (*Umbelliferae*). — Chromosoma (Berl.) **65**, 335—351.

KLEIN, W. H., MURPHY, W., ATTARDI, G., BRITTEN, R. J., DAVIDSON, E. H., 1974: Distribution of repetitive and nonrepetitive sequence transcripts in HeLa mRNA. — Proc. Nat. Acad. Sci. (U.S.) **71**, 1785—1789.

MELLI, M., WHITFELD, C., RAO, K. V., RICHARDSON, M., BISHOP, J. O., 1971: DNA/RNA hybridization in vast DNA excess. — Nature New Biol. **231**, 8—12.

PISTERER, J., KLOPPSTECH, K., 1977: Free and membrane bound polysomes from plant cell cultures. — In: Colloques internat. du CNRS, No. **261**, pp. 279—283. Paris: Edition CNRS.

ROSBASH, H., FORD, P. J., BISHOP, J. O., 1974: Analysis of the C-value paradox by molecular hybridization. — Proc. Nat. Acad. Sci. (U.S.) **71**, 3746—3750.

SEITZ, U., RICHTER, G., 1970: Isolierung und Charakterisierung schnell markierter, hochmolekularer RNA aus freisuspendierten Calluszellen der Petersilie (*Petroselinum sativum*). — Planta (Berl.) **92**, 309—326.

TAUTVYDAS, K. J., 1971: Mass isolation of pea nuclei. — Pl. Physiol. **47**, 499—503.

TIMBERLAKE, W. E., SHUMARD, D. S., 1977: Relationship between nuclear and polysomal RNA populations of *Achlya*: a simple eucaryotic system. — Cell **10**, 623—632.

WALBOT, V., DURE, L. S., 1976: Developmental biochemistry of cotton seed embryogenesis and germination. VII. Characterization of the cotton genome. — J. Molec. Biol. **31**, 349—370.

ZIMMERMANN, J. L., GOLDBERG, R. B., 1977: DNA sequence organization in the genome of *Nicotiana tabacum*. — Chromosoma (Berl.) **59**, 227—252.

Address of the authors: DOROTHEA BARTELS, Dr. MANUEL KIPER, Institut für Botanik der Universität, Herrenhäuserstr. 2, D-3000 Hannover, Federal Republic of Germany.

Pl. Syst. Evol., Suppl. 2, 151—161 (1979)

Institute of Biology II, Department of Genetics,
The University of Tübingen, German Federal Republic

Transcriptional Activity in Seedlings of *Matthiola incana* (*Brassicaceae*) Determined by DNA-RNA Hybridisation

By

Ingrid Gollmer and **Vera Hemleben**, Tübingen

Key Words: Angiosperms, Brassicaceae, *Matthiola incana*. — poly(A)-RNA, DNA-RNA hybridization.

Abstract: From the crucifer *Matthiola incana* the percentage of DNA represented in messenger RNA populations from different stages of development was determined by filter saturation hybridization experiments.—(1) The RNA preparations used were fractionated into poly(A)$^{\ominus}$-RNA and poly(A)-RNA by poly(U)-sepharose chromatography. The size distribution and molecular weights were determined by electrophoresis on polyacrylamide gels in formamide. The average molecular weight of poly(A)-RNA, calculated from the electrophoretic mobility relative to *E. coli* rRNA marker, is about $4 \cdot 10^5$ daltons.—(2) Comparison of the transcription rate of poly(A)-RNA in seedlings, grown either in the light or in the dark, show that poly(A)-RNA synthesis in light-grown seedlings is increased to an extent of about 40% compared with the poly(A)-RNA synthesis in darkgrown seedlings.—(3) The percentage of the genome coding for the various poly(A)-RNA populations and the estimated number of transcribed genes per haploid genome are approximately 0.22% and 6,000 for poly(A)-RNA from light-grown seedlings, and about 0.18% and 4,800 for poly(A)-RNA from dark-grown seedlings. According to the percentage of unique DNA of 30%, the amount of DNA hybridizing to poly(A)-RNA of 0.22% in light-grown seedlings and of 0.18% in dark-grown seedlings correspond to about 0.7% of single-stranded unique DNA (i.e. 1.4% double-stranded unique DNA) and to 0.6% single-stranded unique DNA (i.e. 1.2% double-stranded unique DNA).

The genome of most eukaryotes is of high complexity, and the DNA sequences can be grouped into three general classes based on their frequency in the genome: highly repetitive DNA (also called satellite DNA), middle-repetitive DNA (including the genes for rRNA, tRNA, and histones), and unique DNA (containing structural genes). This complexity makes it difficult to investigate differential gene expression and the control mechanisms involved in the selection of genetic information to be expressed during development. Thus it is advisable to reduce the complexity of the genome by separating the repeated DNA sequences from the informative unique DNA sequences.

One approach to reduce the genome complexity is to reassociate the total denatured DNA to a Cot-value at which the repetitive DNA classes will have hybridized as far as possible to the double-stranded form. Another possibility is to measure the total amount of unique DNA sequences which are active in transcription and, therefore, are represented in the messenger RNA populations at various developmental stages, or in various differentiated tissues—an approach which is used in the experiments described here.

Materials and Methods

Preparation of Nucleic Acids. Germination of seeds of *M. incana* (L.) R. Br. and isolation and purification of DNA was carried out as described by Hemleben&al. (1975). Total DNA was centrifuged on CsCl gradients (8.3 g CsCl dissolved in 6.65 ml Tris-HCl/EDTA, pH 7.5, containing up to 500 μg of DNA; centrifugation at 20 °C for 60 h at 32,000 rpm in a Beckman Ti 50 rotor), except the experiment in Fig. 4, where DNA was centrifuged in an actinomycin D-CsCl gradient (Grierson & Hemleben 1977). After centrifugation, 0.5 ml fractions were collected with an ISCO fractionator.—Radioactive RNA was extracted from *M. incana* seedlings labelled *in vivo* with [5-³H] uridine (27 Ci/mM; 100 μCi/ml or 200 μCi/ml) for different periods of time at various developmental stages. Poly(A)-RNA was purified from total RNA by affinity chromatography on poly(U)-sepharose (Pharmacia) columns.

Gel Electrophoresis. 2.4 % polyacrylamide gels were formed in plexiglass tubes (8 cm long, 6 mm diameter) and electrophoresis of RNA samples was carried out as described by Loening (1969). 4% polyacrylamide gels in formamide were prepared as described by Grierson & Hemleben (1977). Electrophoresis was at 3.5 mA/gel for 3.5–4 h. The gels were then equilibrated in distilled water, scanned on a Joyce-Loebl gel scanner at 254 nm, frozen in dry ice, and sliced into 1 mm sections. Gel slices were incubated in 0.5 ml 12.5 % ammonia solution for 12 h at room temperature, and counted in 3.5 ml Unisolve 1 scintillation mixture.

DNA-RNA Saturation Hybridization. Total DNA samples purified on CsCl gradients or fractions from actinomycin D-CsCl gradients were denatured and attached to nitrocellulose filters (0.45 μ, 13 mm diameter), and the filters were dried at 80 °C in a vacuum oven. Hybridization of filter-bound DNA to the different poly(A)-RNA samples was carried out following the general method of Birnstiel & al. (1968). In the hybridization experiments of "pre-hybridized" DNA to poly(A)-RNA filters were incubated with either unlabelled 25 S and 18 S rRNA or total unlabelled poly(A)⊖-RNA in 6 × SSC + 0.1 % SDS for 3 h at 65 °C, washed several times in SSC solutions, dried, and hybridized to the various poly(A)-RNA samples in 6 × SSC + 0.1%SDS at 67°C for the times indicated.

After hybridization filters were washed in cold 6 × SSC and 2 × SSC, treated with ribonuclease, washed again several times in 2 × SSC, dried, and counted in Toluol-PPO-POPOP scintillator. The amount of DNA present on the filters at the end of the hybridization reaction was measured as described by INGLE & al. (1975). The data obtained were used to carry out a regression analysis.

Voucher specimens of the investigated plants are kept in the herbarium TÜB.

Results and Discussion

Molecular Weight Measurements. In the investigations described here, seedlings of *Matthiola incana* were used and labelled *in vivo* with [5-^{3}H] uridine to a high specific radioactivity at different stages of development. Total cellular RNA was isolated and fractionated by affinity chromatography into non-polyadenylated RNA (poly (A)$^\ominus$-RNA) and polyadenylated RNA (poly(A)-RNA). The total amount of poly(A)-RNA may account for only part of the total messenger RNA since a portion of the mRNA does not contain poly(A)-segments.

The first experiments were carried out in order to determine the molecular weights and size distribution of both poly(A)$^\ominus$-RNA and poly(A)-RNA by gel electrophoresis on polyacrylamide and polyacrylamide in formamide. Fig. 1 *a* shows the electrophoretic fractionation of poly(A)$^\ominus$-RNA on a 2.4% polyacrylamide gel. The two main components represent the ribosomal 25 S RNA and 18 S RNA with the size of $1.3 \cdot 10^6$ daltons and of $0.7 \cdot 10^6$ daltons, respectively. Other components can be detected corresponding in size to the ribosomal RNAs of prokaryotes (23 S and 16 S), which represent the ribosomal RNAs of chloroplasts. The radioactive profile of the same poly(A)$^\ominus$-RNA preparation on a 4% polyacrylamide gel in formamide is shown in Fig. 1 *b*. Formamide was used in order to prevent intermolecular aggregation and to minimize conformational effects due to secondary structures. Already small differences in the molecular weight of RNA molecules can thus be detected. Under the denaturing conditions of formamide the 25 S rRNA is unstable and divides into smaller components.

Fig. 2 shows the electrophoretic mobility of a population of poly(A)-RNA on a 2.4% polyacrylamide gel and on a 4% polyacrylamide gel in formamide. In polyacrylamide the poly(A)-RNA smears throughout the gel from high to rather small molecular weights. In contrast, in formamide the poly(A)-RNA migrates as a rather broad, but distinct peak. Taking the molecular weights of the marker rRNAs of *E. coli* as $1.05 \cdot 10^6$ and $0.52 \cdot 10^6$ daltons (SPOHR & al. 1976), the average size of the poly(A)-RNA is calculated to be approximately $4 \cdot 10^5$ daltons. This molecular weight is possibly an underestimation, since a greater proportion of rather small molecules occurs, suggesting that formamide

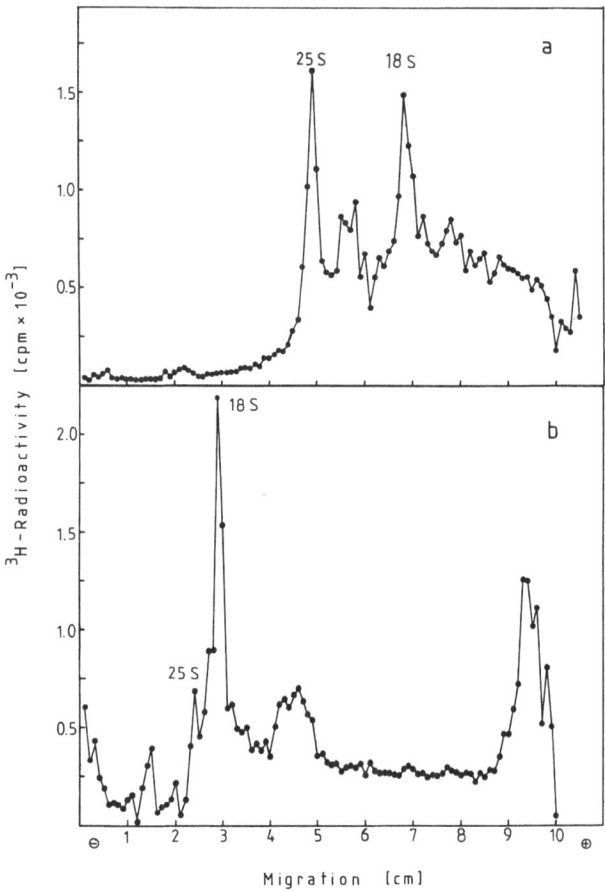

Fig. 1. Gel electrophoresis of poly(A)$^\ominus$-RNA from *M. incana* seedlings labelled with 100 μCi/ml [5-³H]uridine for 24 h from the third to the fourth day of growth: *a* electrophoresis on 2.4% polyacrylamide; *b* electrophoresis on 4.0% polyacrylamide in formamide

treatment causes a fragmentation of the poly(A)-RNA molecules into smaller components. The consequence of this would be that the actual molecular weight of the poly(A)-RNA would be slightly higher than $4 \cdot 10^5$ daltons.

However, this molecular weight of poly(A)-RNA is in good agreement with that found earlier for *M. incana* (GRIERSON & HEMLEBEN 1977) or for sycamore cells (GRIERSON & COVEY 1976), whereas WALBOT &

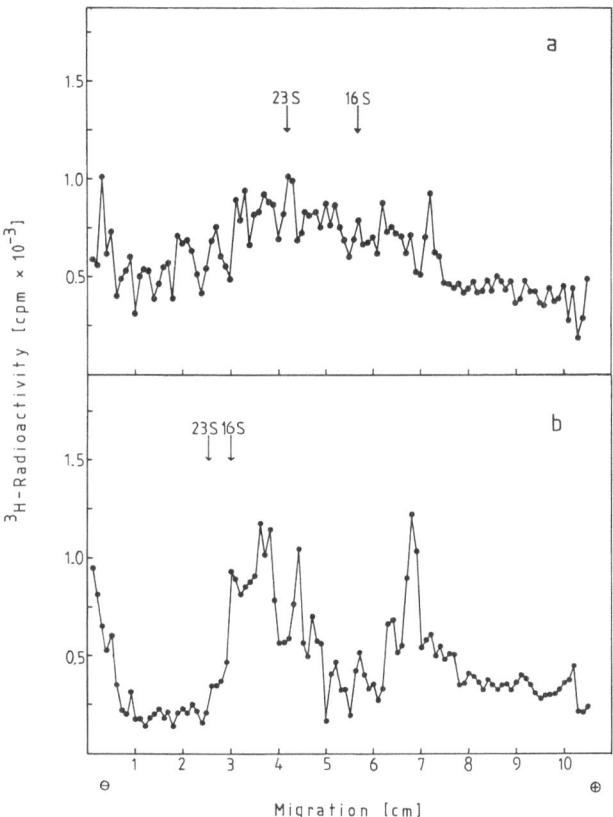

Fig. 2. Gel electrophoresis of poly(A)-RNA from *M. incana* seedlings labelled with 100 μCi/ml [5-³H]uridine for 24 h from the third to the fourth day of growth: *a* electrophoresis on 2.4% polyacrylamide; *b* electrophoresis on 4.0% polyacrylamide in formamide

DURE (1976) reported the molecular weight of poly(A)-RNA from cotton to be about twice this size, but under non-denaturing conditions.

Transcription Rates of Total RNA and Poly(A)-RNA. Comparison of the rate of transcription in seedlings of *M.incana* grown either in the light or in the dark shows that RNA synthesis in seedlings grown in the light is increased to an extent of 30–40% compared with the RNA synthesis in seedlings grown in the dark. This is in accordance with the expectation in view of the increased physiological activity demanded for growth in the light, in contrast to the reduced metabolism under the conditions of

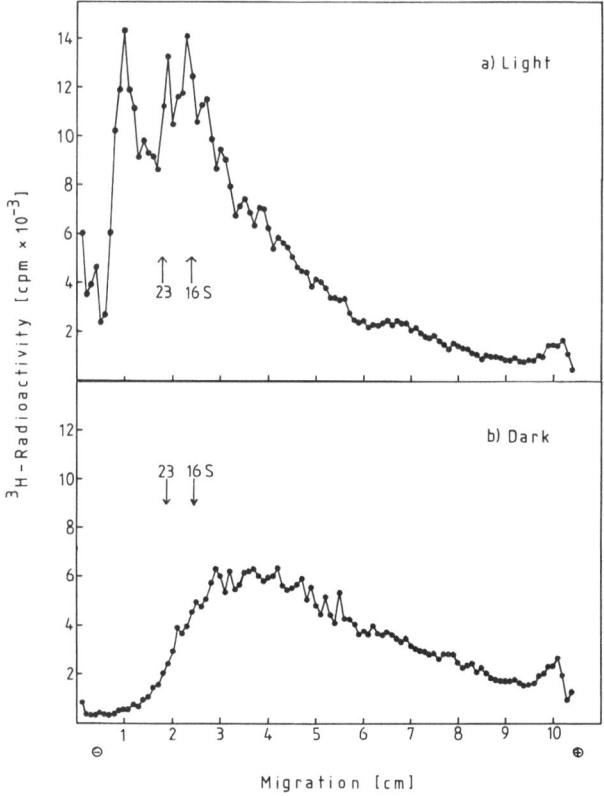

Fig. 3. Gel electrophoresis on 4% polyacrylamide in formamide of poly(A)-RNA from *M. incana* seedlings labelled with 100 μCi/ml [5-³H]uridine for 4 h during the second day of growth: *a* poly(A)-RNA from light-grown seedlings; *b* poly(A)-RNA from dark-grown seedlings

dark-growth. The same can be found for the poly(A)-RNA synthesis in seedlings grown either in the light or in the dark. Fig. 3 shows preparations of poly(A)-RNA from light- and dark-grown seedlings labelled with [5-³H]uridine for 4 h during the second day of development. The amount of poly(A)-RNA synthesized in the light-grown seedlings is about 40 % higher than in dark-grown seedlings.

In this preparation of poly(A)-RNA from light-grown seedlings (Fig. 3 a) some contamination with ribosomal RNA is found. Normally this is only obvious when RNA is labelled for a longer time so that the specific radioactivity of the ribosomal RNA approaches that of the rapidly

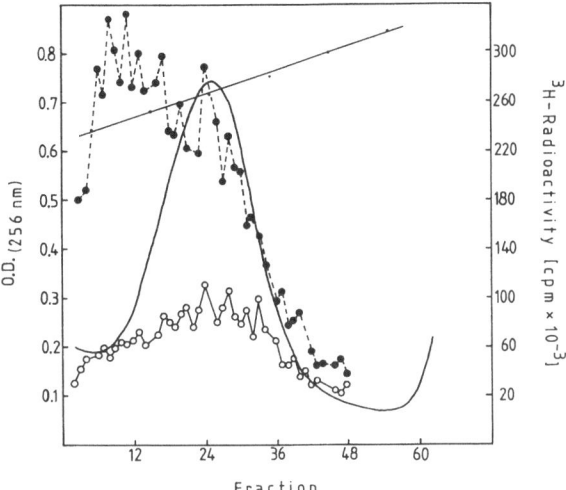

Fig. 4. Hybridization of poly(A)-RNA to DNA fractions from an actinomycin D-CsCl gradient. *M. incana* DNA (300 μg) was centrifuged on an actinomycin D-CsCl gradient, DNA fractions were denatured and loaded onto nitrocellulose filters, and the filters were cut in half. One set of half-filters was hybridized directly to poly(A)-RNA (8 μg/ml; spec. radioact. 52,000 cpm/μg) in 6 × SSC + 0.1% SDS for 72 h at 65 °C (●——●). The other set of half-filters first was incubated with unlabelled 25 S and 18 S rRNA (150 μg/ml) in 6 × SSC + 0.1% SDS for 3 h at 65 °C; filters were then washed and hybridized to poly(A)-RNA (8 μg/ml; spec. radioact. 52,000 cpm/μg) in 6 × SSC + 0.1% SDS for 72 h at 65 °C (○——○). After hybridization filters were washed, treated with ribonuclease, and counted

labelled poly(A)-RNA. Although the extent of contamination by rRNA is not greater than 5-10%, it significantly affects the hybridization experiments as discussed later.

DNA-RNA Saturation Hybridization. The percentage of DNA complementary to the messenger RNA populations from various developmental stages was determined by saturation hybridization with the filter technique. Denatured total DNA bound to nitrocellulose filters was used in saturation hybridization experiments with excess poly(A)-RNA labelled *in vivo* with [5-³H] uridine for 24 h to obtain steady-state labelling conditions. Steady-state labelling was desirable, since RNA of high specific radioactivity was needed. As mentioned earlier, it was not possible to completely remove rRNA contamination under these labelling conditions.

Preliminary hybridization experiments with filter-bound DNA and poly(A)-RNA labelled in this manner showed a relatively high and fast initial hybridization rate followed by a second component hybridizing more slowly. These results suggested the presence of at least two kinetic components being involved in the hybridization reaction, one of which represents contaminating rRNA accountable for the rapid initial hybridization. The experiment in Fig. 4 shows to what extent contaminants of rRNA sequences are capable of affecting the hybridization between DNA and poly(A)-RNA. Total DNA was centrifuged in an actinomycin D-CsCl gradient in order to separate the ribosomal DNA from the main-band DNA (HEMLEBEN & al. 1977). The DNA of every fraction was denatured, loaded onto nitrocellulose filters, and the filters were cut in half. One set of half-filters was hybridized directly to poly(A)-RNA. Two peaks of hybridization can be detected, one peak is coincident with the main-band DNA, and the other one is located on the "light" side of the main-band DNA in the region of the ribosomal DNA, strongly suggesting the presence of contaminating ribosomal RNA molecules in the hybridization mixture. Thus, even little contamination of poly(A)-RNA with ribosomal RNA is shown to result in considerable amounts of hybridization, although it is probably not sufficient to saturate the ribosomal genes. The other set of half-filters first was incubated with unlabelled ribosomal 25 S and 18 S RNA to the saturation level and afterwards hybridized to poly(A)-RNA. Hybridization to poly(A)-RNA now occurs only with sequences of the main band DNA, indicating that this pre-incubation of the filter-bound DNA with ribosomal RNA causes the saturation of the ribosomal genes so that they are not available in the following hybridization reaction with poly(A)-RNA. Therefore, the following hybridization experiments of total DNA and poly(A)-RNA preparations were carried out only after such a "pre-hybridization" of the filters with total unlabelled poly(A)$^{\ominus}$-RNA as separated by poly(U)-sepharose chromatography, containing rRNA, tRNA, and non-polyadenylated mRNA in excess.

Fig. 5 shows the saturation hybridization of "prehybridised" total DNA and poly(A)-RNA from light- and dark-grown seedlings labelled with [5-^{3}H] uridine for 24 h from the third to the fourth day of growth. The hybridization data were used to carry out a regression analysis. The solid lines in Fig. 5 show the regression best fit for a one-component-reaction. The regression results in a saturation value of 0.22 % hybridizing with poly(A)-RNA from light-grown seedlings with a half-time $t_{1/2}$ of 18 h, at which 50 % hybridization has taken place. In the hybridization reaction of total DNA and poly(A)-RNA from dark-grown seedlings the saturation value is calculated as 0.18 % and the half-time $t_{1/2}$ as 21 h.

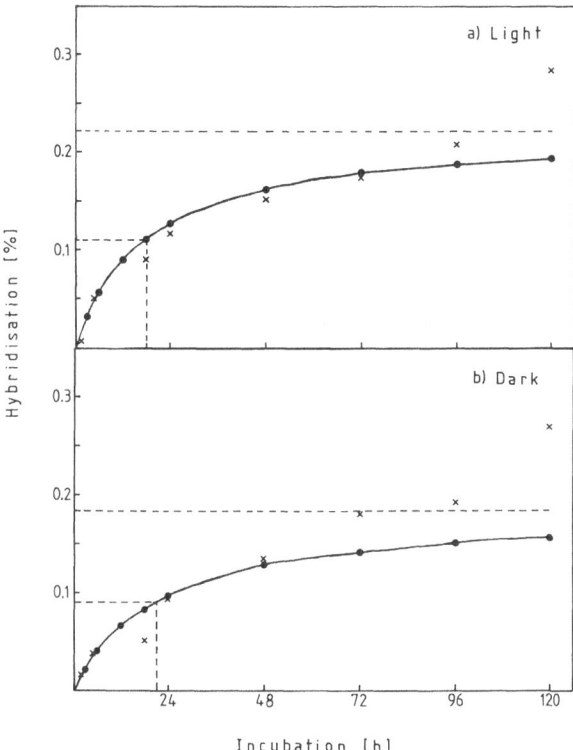

Fig. 5. Hybridization of total DNA to poly(A)-RNA from light- and dark-grown *M. incana* seedlings labelled with 200 μCi/ml [5-³H]uridine for 24 h from the second to the third day of growth. Total DNA purified on CsCl gradients was attached to nitrocellulose filters (20 μg DNA/filter). The filters were pre-incubated with unlabelled poly(A)$^{\ominus}$-RNA (50 μg/ml) in 6 × SSC + 0.1%SDS for 3 h at 65 °C, washed, and hybridized to a) poly(A)-RNA from light-grown seedlings (15 μg/ml; spec. radioact. 2,800 cpm/μg) or to b) poly(A)-RNA from dark-grown seedlings (15 μg/ml; spec. radioact. 2,450 cpm/μg). After hybridization filters were washed, treated with ribonuclease, and counted, followed by the determination of the amount of DNA present on the filters at the end of the hybridization reaction. The solid lines (●————●) show the regression best fit for a one-component reaction of the measured hybridization data (×)

Considering the DNA content of *M. incana* of 1.5 pg per haploid genome and assuming a unique gene to have an average length of about 1,000 nucleotides, the percentage of total DNA of 0.22 % hybridizing to poly(A)-RNA from light-grown seedlings would correspond to an amount of approximately 6,000 genes per haploid genome, which are transcribed at this specific developmental stage; and of about 4,800 transcribed genes, respectively, corresponding to 0.18 % of total DNA hybridizing to poly(A)-RNA from dark-grown seedlings. This number of genes which are possibly active in transcription at that specific stage of development can only refer to the proportion of *polyadenylated* mRNA, which comprises only about 60% of the total cellular mRNA (GRIERSON & COVEY 1976). If the proportion of non-polyadenylated mRNA is assumed to contain specific informative sequences different from that in polyadenylated mRNA, the amount of possibly transcribed genes would be increased by that amount corresponding to the proportion of non-polyadenylated mRNA.

In respect to the amount of unique DNA sequences in *M. incana* which is estimated to be 30 % of total DNA (WENZEL & HEMLEBEN 1979), the hybridization values of 0.22 % (light-grown seedlings) and of 0.18 % (dark-grown seedlings) correspond to about 0.7 % of single-stranded unique DNA, which have hybridized, i.e. 1.4 % of double-stranded unique DNA, and to about 0.6 % of single-stranded unique DNA, i.e. 1.2 % of double-stranded unique DNA.

The model of organisation of unique DNA suggests the length of a unique gene to be about 1,000 nucleotides with interspersions of repeated DNA sequences of about 300–500 nucleotides in length (WENZEL & HEMLEBEN 1979). These estimates and the amount of unique DNA present in the total DNA of *M. incana* would yield an amount of about 600,000 unique genes in total. This would result in a ratio of 1 : 60 of transcribed to present unique genes. However, apart from these speculative calculations it is clear that in specific developmental stages only a very small percentage of the total gene set is transcribed.

References

BIRNSTIEL, M. L., SPEIRS, J., PURDOM, I., JONES, K., LOENING, U. E., 1968: Properties and composition of the isolated ribosomal DNA satellite of *Xenopus laevis*. — Nature **219**, 454—463.
GRIERSON, D., COVEY, S., 1976: The properties and function of rapidly labelled nuclear RNA. — Planta (Berl.) **130**, 317—321.
— HEMLEBEN, V., 1977: Ribonucleic acid from the higher plant *Matthiola incana*. Molecular weight measurements and DNA-RNA hybridization studies. — Biochim. Biophys. Acta **475**, 424—436.
HEMLEBEN, V., ERMISCH, N., KIMMICH, D., LEBER, B., PETER, G., 1975: Studies on the fate of homologous DNA applied to seedlings of *Matthiola incana*. — Eur. J. Biochem. **56**, 403—411.
— GRIERSON, D., DERTMANN, H., 1977: The use of equilibrium centrifugation in actinomycin-caesium chloride for the purification of ribosomal DNA. — Pl. Sci. Lett. **9**, 129—135.

INGLE, J., TIMMIS, J. N., SINCLAIR, J., 1975: The relationship between satellite deoxyribonucleotic acid, ribosomal ribonucleic acid gene redundancy, and genome size in plants. — Pl. Physiol. **55**, 496—501.

LOENING, U. E., 1969: The determination of the molecular weight of ribonucleic acid by polyacrylamide gel electrophoresis. — Biochem. J. **113**, 131—138.

SPOHR, G., MIRAULT, M.-E., IMAIZUMI, T., SCHERRER, K., 1976: Molecular weight determination of animal cell RNA by electrophoresis in formamide under fully denaturing conditions on exponential polyacrylamide gels. — Eur. J. Biochem. **62**, 313—322.

WALBOT, W., DURE, L. S., 1976: Developmental biochemistry of cotton seeds embryogenesis and germination. VII. Characterization of the cotton genome. — J. Molec. Biol. **101**, 503—536.

WENZEL, W., HEMLEBEN, V., 1979: DNA reassociation studies and considerations on the genome organization and evolution of higher plants. — Pl. Syst. Evol., Suppl. **2**, 29—40.

Address of the authors: Dipl.-Biol. INGRID GOLLMER and Prof. Dr. VERA HEMLEBEN, Institut für Biologie II, Lehrstuhl für Genetik der Universität Tübingen, Auf der Morgenstelle 28, D-7400 Tübingen, Federal Republic of Germany.

Pl. Syst. Evol., Suppl. 2, 163—177 (1979)

Institute of Biology I, The University of Tübingen,
German Federal Republic

Synthesis and Release of RNA by Isolated Nuclei of Plant Cells in Relation to Nucleoside Triphosphates and Divalent Cations

By

Klaus Großmann, Hans-Peter Haschke, Ursula Seitz, and
Ulrich Seitz, Tübingen

Key Words: Angiosperms, *Petroselinum crispum.* — Isolated nuclei, *in vitro* transcription, RNA release.

Abstract: An *in vitro* system for nuclei from freely suspended callus cells from *Petroselinum crispum* is described. A filtration technique allowed the measurement of transcription and release of RNA simultaneously. Both can be stimulated by nucleoside triphosphates and divalent cations in equimolar concentrations. Divalent cations have an inhibitory effect on the release of RNA. Nucleoside triphosphates, by means of their complex-forming capacity, can remove divalent cations, and so stimulate RNA translocation. Sodiumpyrophosphate also forms chelates with divalent cations and increases RNA release. Apparently, nucleoside triphosphates are used as a complex in translocating RNA.—The stimulatory effect on RNA synthesis results from increasing complex concentrations formed by nucleoside triphosphates and divalent cations, indicating that the complex could be the substrate of the RNA polymerases. The experiments with sodiumpyrophosphate show that the chelating effect could also be a part of the process. With ADP RNA synthesis can be inhibited.—Isolated nuclei exhibit a powerful NTPase and pyrophosphatase activity, whereas ADP is not split by the nuclei.—The effects of ATP, sodiumpyrophosphate and ADP on maintaining or reducing RNA synthesis should be seen in connection with the enzymatic splitting of these compounds by various phosphatases (pyrophosphatase, NTPase).

In eukaryotic cells, the regulation of the genome-dependent metabolism is very different when compared with prokaryotes. The most fundamental difference in cell organization is the strict compartmentation between nucleus and cytoplasm, and thus a spatial separation of transcription and translation, in eukaryotes. Additional possibilities of regulation are, among others: 1) multiple DNA-dependent RNA-polymerases with specific properties, 2) regulation at the level of precursor molecules for rRNA and mRNA, and 3) regulation of the transport of ribonucleoprotein particles from the nucleus into the cytoplasm. For investigations of these processes, *in vitro* systems with isolated nuclei are of particular advantage.

In this communication we describe such an *in vitro* system utilizing isolated plant nuclei from a suspension culture of parsley. We looked especially for the role of nucleoside triphosphates and divalent cations in transcription and release of RNA. It has been demonstrated that ATP is able to stimulate the release of RNA in various systems of nuclei isolated from animal cells (SCHNEIDER 1959, ISHIKAWA & al. 1969, CHATTERJEE & WEISSBACH 1973). On the other hand, the release is inhibited by divalent cations, especially Mg^{2+}. It is believed that ATP is not only used as a mere energy source, but also acts as a chelating agent and removes Mg^{2+} by its complex-forming capacity (SCHNEIDER 1959, ISHIKAWA & al. 1969, CHATTERJEE & WEISSBACH 1973, SAUERMANN 1976).

The involvement of an enzyme, which hydrolyses nucleoside triphosphates in the translocation process from the nucleus to the cytoplasm, was demonstrated in nuclei from fibroblasts (AGUTTER & al. 1976). Such an enzyme seems to be located in the nuclear pore complex (FRANKE 1974).

The situation in cells of higher plants is still badly understood. Especially in the case of transcription, hardly anything is known about the interaction of divalent cations and nucleoside triphosphates.

In the present report we describe experiments with isolated nuclei from plant cells which shed some light on the role of divalent cations and nucleoside triphosphates in the release and the synthesis of RNA.

Materials and Methods

Plant Material. Cells of *Petroselinum crispum* (MILL.) A. W. HILL were propagated as previously described (GEBAUER & al. 1975).

Solutions. Isolation medium 1 (IM 1) contained 20 mM Tris · HCl (pH 7.8), 200 mM sucrose, 2 mM $CaCl_2$, 10 mM 2-mercaptoethanol, 5 mM $MgCl_2$, 10 mM KCl, 2.5 % Ficoll (w/v) and 5 % Dextran T 40 (w/v). Isolation medium 2 (IM 2) was similar, but 0.02 % Triton X-100 were added. Isolation medium 3 (IM 3) equalled IM 1, but lacked sucrose.

Preparation and Purification of Nuclei. Nuclei were isolated by a modification of the methods of TAUTVYDAS (1971) and GEBAUER & al. (1975) and purified by a modified method of CHEN & al. (1975). Cells (fresh weight 50 g) were incubated in IM 1 containing 2 % pectinase (Serva, Heidelberg) and 1.5 % cellulase (Onozuka R 10, Yakult Biochemicals, Japan) at 30 °C for 4 h. The resulting crude protoplasts were collected on a 100 μm nylon net and brought to a final volume of 160 ml with IM 1. The suspension was homogenised in a Potter-Elvehjem homogeniser. After filtration through a series of nylon nets the suspension was centrifuged. The pellet was resuspended in IM 2 and sedimented once more. The nuclear pellet was resuspended in IM 1, layered on a sucrose cushion (23 ml IM 3 containing 1.2. M sucrose) and centrifuged at 20,000 rpm for 7 min at 2 °C in a Beckman centrifuge using a SW 25.1 rotor. The nuclear pellet was now resuspended in IM 3 containing 1.0 M sucrose, centrifuged through a discontinuous gradient consisting of 10 ml each of 1.2, 1.4, 1.8 M sucrose for 10 min at 4,000 rpm using a SW 25.1 rotor. The purified nuclei were collected from the 1.2 and 1.4 M zones. The nuclei could be stored at —20 °C for 8 days without losing their activity if 40 % (v/v) glycerol was present. A 10 min centrifugation at 1,200 g freed them from the storage medium. For measurements of transcription the nuclei were suspended in IM 1 at a concentration of 8.10^6 nuclei/ml.

RNA Synthesis in Isolated Nuclei. The complete incubation medium modified from HAMILTON & al. (1972), kept in ice, contained 20 mM Tris · HCl (pH 7.8), 200 mM sucrose, 10 mM 2-mercaptoethanol, 1 mM $MnCl_2$, 5 mM $MgCl_2$, 10 mM KCl, 1 mM $CaCl_2$, 1.25 % Ficoll, 2.5 % Dextran T 40, 1 mM each of ATP, GTP, CTP, UTP, 0.3 μCi [5-³H] uridine 5'-triphosphate (Amersham Buchler, Braunschweig) and 0.25 ml of nuclei suspension in IM 1 in a total volume of 0.5 ml, added last in order to start the transcription reaction. The temperature of the incubation was 36 °C.—Transcription was stopped by adding 0.1 mg bovine serum albumin and 0.1 M sodium pyrophosphate ($Na_4P_2O_7$) dissolved in 0.5 ml of double destilled water.—The nuclear fraction was collected on Millipore filters (RA 1.2 μm, ⌀ 25 mm).

The filter discs bearing the nuclei were transferred to a filter apparatus and treated with 20% TCA (in 0.01 M $Na_4P_2O_7$). To the filtrate which dripped into 1 ml of 20 % TCA dissolved in 0.01 M $Na_4P_2O_7$ 3 ml of 30 % TCA (in 0.01 M $Na_4P_2O_7$) were added. The RNA was allowed to precipitate for 1-2 h at 4 °C. The TCA insoluble material of the filtrate was also collected on Millipore filters (see above). In both cases, the filters were washed four times with 2 % TCA and once with 96 % ethanol at —20 °C.—After drying (60 °C) the activity of the filter discs was counted in toluene-PPO-POPOP.

Phosphatase Activity was measured by determination of P_i released according to LIN & MORALES (1977).—The determination of NTPase activity took place in reagent tubes which contained 0.96 ml of 50 mM Tris · HCl buffer pH 7.8. The reaction was started by quickly adding of 20 μl concentrated substrate solution (as indicated) and 20 μl nuclei suspension ($1 · 10^5$ nuclei). The solution was incubated in a water bath at 36 °C for 10 min. The reaction was stopped by adding of 1 ml molybdovanadate reagent. The absorbance of the color complex, formed by P_i and molybdovanadate, was measured after 3 min at 366 nm in an Eppendorf spectrophotometer. The P_i concentration in the colored solution was measured by comparison with the calibration curve using Na_2HPO_4 solution of known concentrations as standards and substraction of the blank assay without nuclei.

Results

Nuclei prepared from cells of a suspension culture of *Petroselinum crispum* were incubated in a cell free system. The RNA synthesis was measured by the incorporation of [³H] UMP into TCA-insoluble

166 K. GROSSMANN et al.:

material. We were able to determine simultaneously the RNA transcribed, and the RNA released by the nuclei, with the filtration technique used in this investigation. In the present report we were interested in the influence of nucleoside triphosphates and divalent

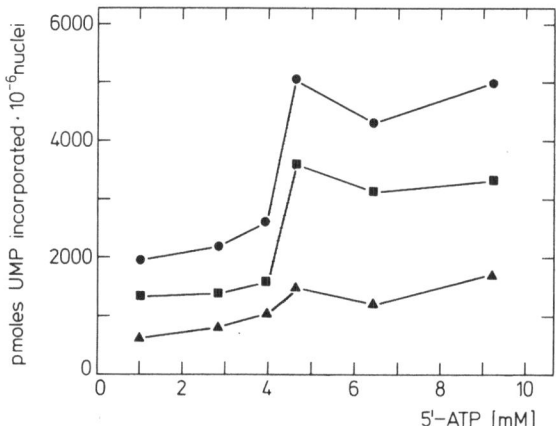

Fig. 1. ATP dependence of RNA synthesis and release in isolated nuclei from *Petroselinum crispum*. Isolation of nuclei, conditions for incubation and radioactivity determination are described in "Materials and Methods". The nuclear suspension was added to the incubation mixture to start the reaction. Temperature was 36 °C and the incubation time was 5 min. The following constituents of the incubation medium were kept constant: 1 mM MnCl$_2$, 1 mM CaCl$_2$, 5 mM MgCl$_2$ (in total 7 mM), 1 mM each of UTP, GTP, CTP; the other conditions were the same as in the standard assay. The ATP concentration was varied from 1-9 mM. — ■———■ incorporation of [³H] UMP into RNA of the nuclear fraction (RNA transcribed but not released) ▲———▲ TCA insoluble material in the filtrate, separated by filtration from the nuclear fraction (see Materials and Methods), representing the RNA transcribed and released from the nuclei (RNA transport). ●———● the total [³H]UMP incorporated, calculated by addition of the two other curves

cations on synthesis and release of RNA. Both processes were stimulated by increasing ATP concentrations in presence of 3 mM other nucleoside triphosphates (at each 1 mM GTP, CTP, UTP) and 7 mM divalent cations, if the concentration of nucleosides was equimolar to the divalent cations (Fig. 1). The elucidation of this phenomenon was the subject of further experiments. RNA transport and RNA synthesis are considered separately.

Influence of Divalent Cations and Nucleoside Triphosphates on RNA Release. The control of the release of RNA from nuclei by divalent

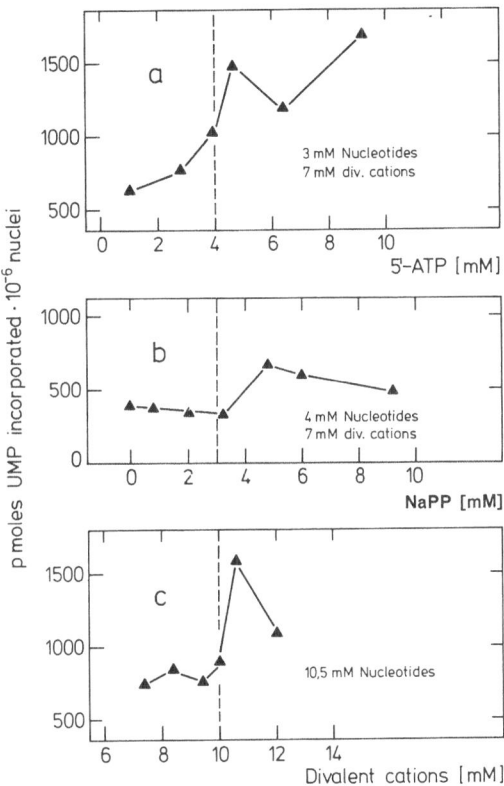

Fig. 2. RNA transport under various conditions. — *a* ATP dependence of RNA transport. The following constituents of the incubation medium were kept constant. 1 mM $MnCl_2$, 1 mM $CaCl_2$, 5 mM $MgCl_2$ (in total 7 mM); 1 mM each of UTP, GTP, CTP. The ATP concentration was varied from 1—9 mM. — *b* Sodiumpyrophosphate influence on RNA transport. The following constituents were kept constant: 7 mM divalent cations (see *a*), 1 mM each of ATP, GTP, UTP, CTP (in total 4 mM). The sodiumpyrophosphate concentration was varied from 0—9 mM. — *c* Influence of Mg^{2+}-ions on the RNA transport. The following constituents were kept constant: 1 mM each of GTP, UTP, CTP, 7.5 mM ATP (in total 10.5 mM); 1 mM $MnCl_2$, 1 mM $CaCl_2$. The Mg^{2+} concentration was varied from 5—10 mM. — For incubation conditions see Fig. 1. The broken line marks a nucleotide (plus sodiumpyrophosphate in *b*): divalent cation ratio of 1

cations and nucleoside triphosphates has been studied intensively in animal cells. In plant cells near to nothing is known about the influence of these components on the release of RNA. The data from Fig. 1 and 2 *a* clearly demonstrate that the release was stimulated, if equimolarity was reached between nucleoside triphosphates and divalent cations. Free ATP (not complexed by divalent cations) added in excess could maintain the RNA release at this higher level. If ATP was partly replaced by sodiumpyrophosphate (Fig. 2 b), the release was also

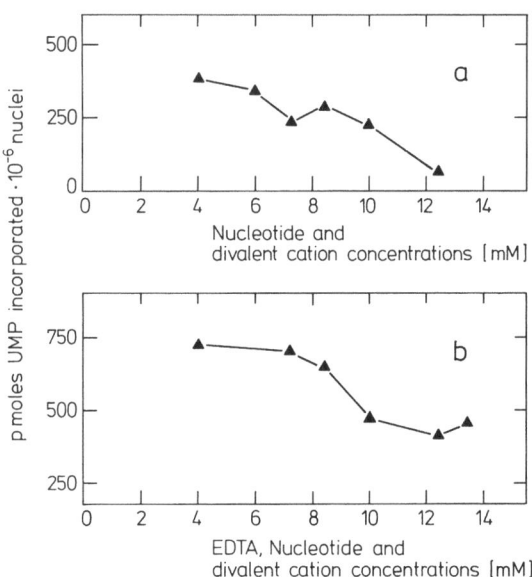

Fig. 3. Influence of complex concentration on RNA transport. — a Complex dependence (divalent cations and nucleotides in ratio 1) of RNA transport. The following constituents were kept constant: 0.5 mM each of GTP, UTP, CTP; 2.5 mM ATP (in total 4 mM); ATP and $MgCl_2$ concentrations were varied, while maintaining a Mg^{2+} : ATP ratio of 1 (in total a nucleotide: divalent cation ratio of 1). — b Complex dependence (divalent cations: nucleotides plus EDTA in ratio 1) of RNA transport. The following constituents were kept constant: 0.5 mM each of GTP, UTP, CTP, ATP; 2 mM EDTA (in total 4 mM); 0.5 mM $MnCl_2$, 1 mM $CaCl_2$, 2.5 mM $MgCl_2$ (in total 4 mM). EDTA and $MgCl_2$ concentrations were varied, while maintaining a nucleotide plus EDTA: divalent cation ratio of 1. — For incubation conditions see Fig. 1

stimulated with increasing concentrations of sodiumpyrophosphate in the presence of 4 mM nucleoside triphosphates (1 mM each of ATP, GTP, UTP, CTP) and 7 mM divalent cations. In this case the concentrations of sodiumpyrophosphate and nucleoside triphosphates together were equivalent to the concentration of divalent cations. At higher concentrations of sodiumpyrophosphate, the RNA release did decline.

This means that ATP could partially be replaced by a complex forming agent which binds divalent cations preventing their inhibitory effect on the release of RNA.

RNA release in the presence of 10.5 mM nucleoside triphosphates (7.5 mM ATP, 1 mM each of GTP, UTP, CTP) was stimulated with increasing divalent cation concentrations, especially $MgCl_2$ (1 mM $CaCl_2$, 1 mM $MnCl_2$) if the ratio of divalent cations: nucleotides was about 1. If the ratio was less than or greater than 1, the release was depressed (Fig. 2 c).

If the concentration of nucleoside triphosphates and divalent cations was increased simultaneously, thus maintaining a cation: nucleotide ratio of 1:1, the release decreased (Fig. 3 a), even if ATP was partially replaced by a complex-forming agent such as EDTA (Fig. 3 b). These experiments demonstrate that maximal RNA release was reached, if divalent cations and nucleoside triphosphates were present in an 1:1 ratio. With increasing concentrations of both components, while maintaining the 1:1 ratio, the RNA release was inhibited. The results also indicate that free divalent cations (not bound in a complex by nucleoside triphosphates or other complex-forming agents) reduced the release of RNA, whereas free nucleoside triphosphates were able to maintain the stimulatory effect.

Influence of Divalent Cations and Nucleoside Triphosphates on RNA Synthesis. The RNA synthesis like the RNA release was stimulated with increasing concentrations of ATP in presence of 3 mM other nucleoside triphosphates (1 mM each of GTP, UTP, CTP) and 7 mM divalent cations (5 mM $MgCl_2$, 1 mM each of $CaCl_2$, $MnCl_2$), if equimolarity was reached between nucleoside triphosphates and divalent cations. Free ATP in excess could also maintain the elevated level of RNA synthesis (Fig. 1, 4 a).

ATP partially replaced by sodiumpyrophosphate produced the same stimulatory effect, if sodiumpyrophosphate plus nucleoside triphosphates and divalent cations were present in an 1:1 ratio (Fig. 4 b). Sodiumpyrophosphate added in excess was also able to maintain the stimulation but to a lower extent. In this case the stimulation was only 46% compared with ATP (Table 1).

In an experiment with ADP instead of ATP in presence of 4 mM nucleoside triphosphates (1 mM each of ATP, GTP, CTP, UTP) and 7 mM divalent cations (Fig. 4c), a decrease in RNA synthesis was observed, if the concentration of ADP exceeded the 1:1 ratio of divalent cations and nucleoside triphosphates plus ADP. To explain such an effect an experiment was designed in which nucleoside triphosphates and divalent cations were increased simultaneously maintaining a cation:

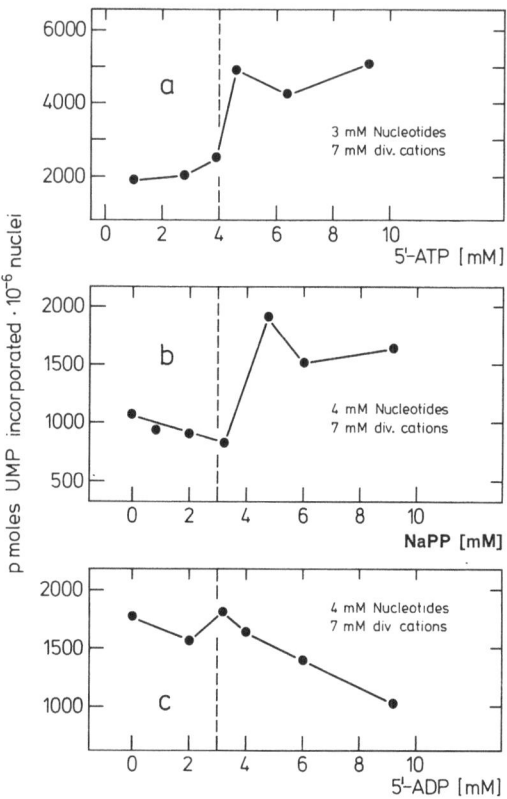

Fig. 4. RNA synthesis under various conditions (the radioactivity of the total TCA insoluble fraction is shown). — a ATP dependence of RNA synthesis. The following constituents were kept constant: 1 mM MnCl₂, 1 mM CaCl₂, 5 mM MgCl₂ (in total 7 mM); 1 mM each of UTP, GTP, CTP. The ATP concentration was varied from 1—9 mM. — b Sodiumpyrophosphate influence on RNA synthesis. The following constituents were kept constant: 7 mM divalent cations (see above); 1 mM each of ATP, UTP, GTP, CTP (in total 4 mM). The sodiumpyrophosphate concentration was varied from 0—9 mM. — c ADP influence on RNA synthesis. The following constituents were kept constant: 1 mM each of ATP, GTP, UTP, CTP (in total 4 mM); 7 mM divalent cations (see above). The ADP concentration was varied from 0—9 mM. — For incubation conditions see Fig. 1 and section Materials and Methods. The broken line marks a nucleotide (plus sodiumpyrophosphate as indicated): divalent cation ratio of 1

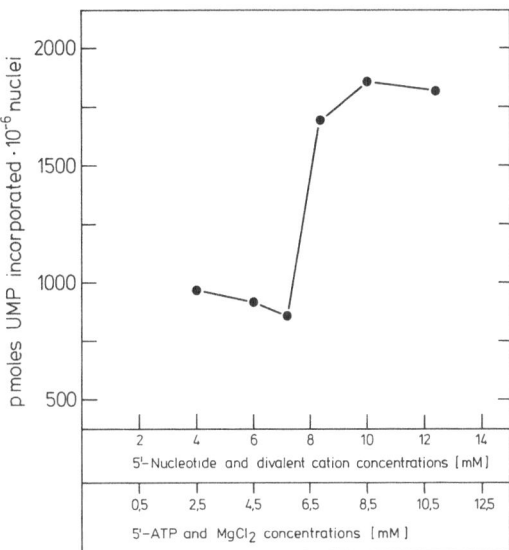

Fig. 5. Influence of complex concentration on RNA synthesis (divalent cations: nucleotides in ratio 1). The radioactivity of the total TCA insoluble fraction is shown. The following constituents were kept constant: 0.5 mM each of GTP, UTP; CTP; 2.5 mM ATP (in total 4 mM); 0.5 mM $MnCl_2$, 1 mM $CaCl_2$, 2.5 mM $MgCl_2$ (in total 4 mM). ATP and $MgCl_2$ concentrations were varied, while maintaining a Mg^{2+}: ATP ratio of 1 (in total a nucleotide: divalent cation ratio of 1). For incubation conditions see Fig. 1

Table 1. Stimulatory effect of 5'ATP and sodiumpyrophosphate (NaPP). The total p moles of UMP incorporated by 10^6 nuclei were determined and compared with one another. For experimental conditions see legends to Figs. 4 a and b

Conditions	p moles UMP incorporated by 10^6 nuclei	Stimulation p moles	%
NaPP: 3.2 mM	823.3		
4.8 mM	1,929.8	1,106.5	46
ATP: 3.9 mM	2,665.8		
4.6 mM	5,087.7	2,421.9	100

nucleoside triphosphate ratio of 1. Fig. 5 shows that the RNA synthesis was stimulated with increasing complex concentrations, formed by nucleoside triphosphates and divalent cations. So we are able to interpret the stimulatory effect of increasing ATP concentrations (shown in Figs. 1 and 4 a) by increasing complex concentrations formed by nucleoside triphosphates and divalent cations. However, also the complex-forming agent sodiumpyrophosphate stimulated RNA synthesis, thus indicating that the complex-forming capacity could play a part in the stimulation of RNA synthesis.—The opposite effect was

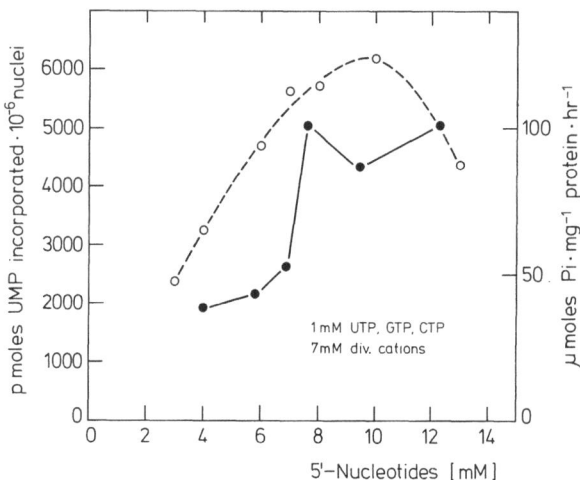

Fig. 6. Nucleotide, especially ATP, dependence of RNA synthesis and nucleosid triphosphatase activity — a comparison. The following constituents were kept constant: 1 mM each of GTP, UTP, CTP; for RNA synthesis 1 mM MnCl$_2$, 1 mM CaCl$_2$, 5 mM MgCl$_2$ (in total 7 mM); for NTPase activity determination 1 mM CaCl$_2$, 6 mM MgCl$_2$ (in total 7 mM). The ATP concentration was varied from 1—9 mM. ●————● The radioactivity of the total TCA insoluble fraction is shown (RNA synthesis). For incubation conditions see Fig. 1. — ○————○ NTPase activity measured by determination of P$_i$ released

observed, if increasing ADP concentrations instead of ATP were applied. ADP concentrations exceeding the ratio of divalent cations: nucleoside triphosphates plus ADP of 1:1 had an inhibitory effect on RNA synthesis. The results described here led to the question, whether other enzymes besides the RNA polymerases, for example phosphatases, could participate in the stimulation of RNA synthesis.

In Fig. 6, the RNA synthesis and the NTPase activity of the isolated nuclei were compared. The NTPase activity reached its maximum at a nucleoside triphosphate: divalent cation ratio of about 1:1. In Fig. 7 a the stimulation of RNA synthesis with increasing complex con-

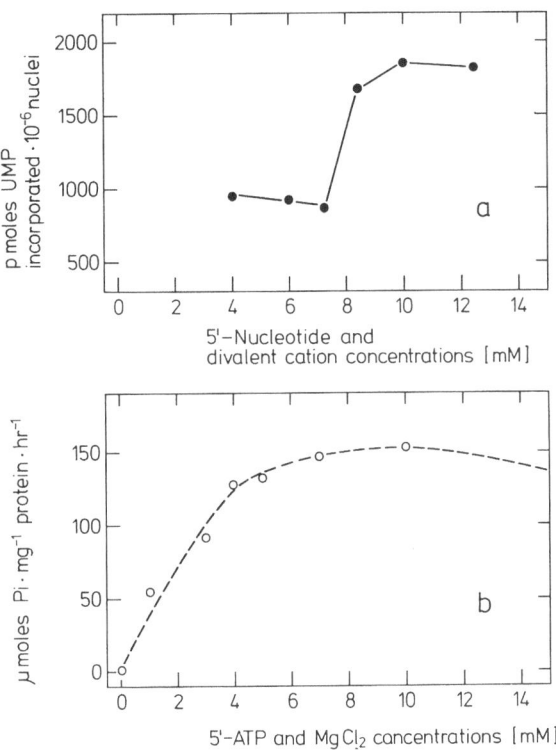

Fig. 7. Influence of complex concentration on RNA synthesis and ATPase activity — a comparison (divalent cations: nucleotides in ratio 1). — *a* Influence of complex concentration on RNA synthesis. The radioactivity of the total TCA insoluble fraction is shown. The following constituents of the incubation medium were kept constant: 0.5 mM each of GTP, UTP, CTP; 2.5 mM ATP (in total 4 mM); 0.5 mM MnCl$_2$, 1 mM CaCl$_2$, 2.5 mM MgCl$_2$ (in total 4 mM). ATP and MgCl$_2$ concentrations were varied, while maintaining a Mg^{2+}: ATP ratio 1 (in total a nucleotide: divalent catio ratio of 1). For incubation conditions see Fig. 1. — *b* Influence of complex concentration on ATPase activity. The ATPase activity measured by determination of P$_i$ released is shown. ATP and MgCl$_2$ concentrations were varied, while maintaining a Mg^{2+}: ATP ratio 1. For incubation conditions and determination of the activity see section Materials and Methods.

centrations is shown. It was compared with the ATPase activity (Fig. 7 *b*) under increasing complex concentrations (MgCl$_2$ and ATP in the ratio of 1:1. The important fact was that NTPases, especially the ATPase, had their V_{max} within the same range of nucleoside triphosphate and divalent cation concentrations, in which the stimulation of RNA synthesis took place.

It would be a reasonable assumption that the NTPase activity was in some way involved in RNA synthesis. Moreover, the pyrophosphatase and ADPase activities were measured in isolated nuclei (Table 2). The nuclei contained a very powerful pyrophosphatase activity in addition to the NTPase activity. ADP was not hydrolyzed by the nuclei.—These results could indicate that various effects of ATP, sodium pyrophosphate

Table 2. Substrate dependence of phosphatase activity of isolated nuclei measured by determination of P_i released

Substrate (7 mM)	Phosphatase activity (μmoles $P_i \cdot mg^{-1}$ protein $\cdot hr^{-1}$)
ATP·Mg	103.8 ± 11.2
GTP·Mg	121.6 ± 18.3
UTP·Mg	71.6 ± 3.9
CTP·Mg	118.1 ± 1.3
ADP·Mg	Zero
GDP·Mg	51.5 ± 2.5
UDP·Mg	71.4 ± 2.9
CDP·Mg	110.5 ± 8.3
pNP·Mg	214.1 ± 5.5
NaPP·Mg	155.4 ± 6.9
ATP·Mg	113.9 ± 2.5
ATP·Ca	102.4 ± 1.7
ATP·Mn	104.5 ± 4.4
ATP	118.9 ± 5.6

and ADP on RNA synthesis could be connected with the enzymatic splitting of these compounds by phosphatases such as NTPase or pyrophosphatase.

Discussion

In the present study we describe the use of isolated nuclei from freely suspended callus cells of parsley for studying transcription *in vitro*. The determination of radioactivity incorporated into TCA-insoluble material by a filtration technique enabled us to examine both transcription and release of newly synthesized RNA simultaneously. Data were obtained about the influence of divalent cations and nucleoside triphosphates on synthesis and release of RNA. Experiments with nuclei from animal cells showed an antagonistic effect of nucleoside triphosphates, especially ATP, and divalent cations, especially Mg^{2+}, on the release of RNA (SCHNEIDER 1959, ISHIKAWA & al. 1969, CHATTERJEE & WEISSBACH 1973, SAUERMANN 1976). ATP might act as a chelating agent removing Mg^{2+}-ions, which had an inhibitory effect on the release of RNA. It is believed that the translocation of RNA can be regulated by the balance of these two components. On the other hand, data are available that ATP and the other nucleoside triphosphates were substrates for a nucleoside triphosphatase, located in the pore complex of the nuclear envelope (AGUTTER & al. 1976, FRANKE 1974). It is now evident that the NTPase is involved in the nucleocytoplasmic translocation of ribonucleoproteins. In plant cells examined here, the release of newly synthesized RNA was stimulated, if equimolarity between nucleoside triphosphates and divalent cations was reached. The surplus of nucleoside triphosphates, especially ATP, could maintain the elevated level of RNA release; the surplus of divalent cations, especially Mg^{2+}, led to an inhibition of RNA release. Since sodiumpyrophosphate in combination with divalent cations was also able to stimulate the RNA release, it is reasonable to suggest that, in plant cells, the translocation of RNA is also controlled by an antagonistic effect of nucleoside triphosphates and divalent cations. One part of the process is the complex-forming property of the nucleoside triphosphates by binding divalent cations which had an inhibitory effect on the translocation process. On the other hand, it would be also a reasonable explanation that nucleoside triphosphates and divalent cations were utilized together in complex form as substrate in the translocation process, while increasing complex concentration would lead to an inhibition of RNA release.

In the case of RNA synthesis, it is well known that nucleoside triphosphates are used as precursors for RNA, and that divalent cations, especially Mg^{2+} and Mn^{2+}, are necessary for the activity of the RNA-polymerases, but little is known about the interaction of both components in transcription. In the experiments with plant nuclei described here, both processes, RNA release and RNA synthesis, were stimulated by a situation in which nucleoside triphosphates and divalent cations were present in equimolar concentrations (ATP applied in excess of the ratio of about 1 : 1 was able to maintain the higher level in RNA synthesis which was reached after stimulation). The stimulation of RNA synthesis resulting from increasing complex concentrations of nucleoside triphosphates and divalent cations indicates that the complex could be the substrate of the RNA polymerases.

Using inorganic pyrophosphate instead of ATP, stimulation of RNA synthesis was only 46 %, thus indicating that the chelate-forming effect could also be a part of the stimulatory process.

The enzyme activity of NTPase, especially ATPase, had its V_{max} at a substrate concentration, at which the stimulation of RNA synthesis by ATP reached its maximum. It seems, therefore, possible that NTPase is involved in RNA synthesis. Moreover, the nuclei contained a high pyrophosphatase activity, and ADP was not hydrolyzed by the isolated nuclei.

SIEBERT (1968) advanced a hypothesis that, in isolated animal nuclei, the NTPase activity could participate in the control of precursor pools for nucleic acid synthesis. LIBANTI & TANDLER (1969) suggested that high levels of inorganic orthophospate, found in animal nuclei as well as in plant nuclei, could affect the association and dissociation of nucleic acids with proteins and could be involved in the activation of repressed chromatin. Considering the results presented here, we put forward the following hypothesis: A pool of inorganic orthophosphate anions in the nucleus is controlled by the activity of phosphatases. Accordingly an increase of orthophosphates by hydrolyzing, for example, nucleoside triphosphates or pyrophosphate, would activate repressed chromatin and stimulate transcription. Consequently ADP, which could not be hydrolyzed, was not able to enrich the orthophosphate pools of the nuclei and to activate repressed chromatin. Moreover, at increasing ADP concentrations, RNA synthesis is inhibited. Further experiments are necessary to prove or disprove our hypothesis.

References

AGUTTER, P. S., McARDUE, H. J., McCALDIN, B., 1976: Evidence for involvement of nuclear envelope nucleoside triphosphatase in nucleocytoplasmic translocation of ribonucleoprotein. — Nature **263**, 165—167.

CHATTERJEE, N. K., WEISSBACH, H., 1973: Release of RNA from HeLa cell nuclei. — Arch. Biochem. Biophys. **157**, 160—167.

CHEN, Y.-M., LIN, C.-Y., CHANG, H., GUILFOYLE, T. J., KEY, J. L., 1975: Isolation and properties of nuclei from control and auxin-treated soybean hypocotyl. — Pl. Physiol. **56**, 78—82.

FRANKE, W. W., 1974: Structure, biochemistry, and functions of the nuclear envelope. — Int. Rev. Cytol. Suppl. **4**, 71—263.

GEBAUER, H. U., SEITZ, U., SEITZ, U., 1975: Transport and processing of ribosomal RNA in plant cells after treatment with cycloheximide. — Z. Naturforsch. **30 c**, 213—218.

HAMILTON, R. H., KUNSCH, U., TEMPERLI, A., 1972: Simple rapid procedures for isolation of tobacco leaf nuclei. — Analyt. Biochem. **49**, 48—57.

ISHIKAWA, K., KURODA, C., OGATA, K., 1969: Release of ribonucleoprotein particles containing rapidly labelled ribonucleic acid from rat liver nuclei. Effect of adenosine 5'-triphosphate and some properties of the particles. — Biochim. Biophys. Acta **179**, 316—331.

LIBANTI, C. M., TANDLER, C. J., 1969: The distribution of the water-soluble inorganic orthophosphate within the cell: Accumulation in the nucleus. — J. Cell Biol. **42**, 754—765.

LIN, T., MORALES, M. F., 1977: Application of a one-step procedure for measuring inorganic phosphate in the presence of proteins. The actomyosin ATPase system. — Analyt. Biochem. **77**, 10—17.

SAUERMANN, G., 1976: Studies on ribonucleic acid metabolism using nuclear columns. Release of rapidly labelled RNA from rat liver nuclei. — Hoppe-Seylers Z. Physiol. Chem. **357**, 1117—1124.

SCHNEIDER, J. H., 1959: Factors affecting the release of nuclear ribonucleic acid from the nucleus *in vitro*. — J. Biol. Chem. **234**, 2728—2732.

SIEBERT, G., 1963: Enzyme of cancer nuclei. — Exp. Cell Res. Suppl. **9**, 389—417.

— 1968: Nucleus. — In FLORKIN, M., STOTZ, E. H., (Eds.): Comprehensive Biochemistry **23**, 1—37. — Amsterdam-New York-London: Elsevier Publishing Company.

Address of the authors: Dipl.-Biol. KLAUS GROSSMANN, Dipl.-Biol. HANS-PETER HASCHKE, Dr. URSULA SEITZ, and Dr. ULRICH SEITZ, Institut für Biologie I der Universität Tübingen, Auf der Morgenstelle 1, D-7400 Tübingen, Federal Republic of Germany.

Pl. Syst. Evol., Suppl. 2, 179—184 (1979)

Institute of Human Genetics, The University
of the Saarland, Federal Republic of Germany

Does a Compensational Mechanism Exist Among the Nucleolus Organizing Regions (NORs)?

By

Heinrich Zankl and Hanno Huwer, Homburg/Saar

Key Words: *Homo.* — Nucleolus organizer, silver staining, gene dosis compensation.

Abstract: From the fact that the number of NORs shows an extreme variation between the individuals of one species, it is concluded that a mechanism should exist which ensures the supply with ribonucleoproteins in the cell. Reports are reviewed dealing with this subject and two original studies are presented.—The number of satellite associations, and NOR-silver staining, were studied in a human female with 13/13 Robertsonian translocation and her relatives. The proband showed a higher association frequency and a more intensive silver staining of chromosome no. 22 when compared to her parents and her sister. It seems likely that this difference was caused by the loss of two NORs after the fusion of both chromosomes no. 13.—In a second study, 22-monosomic mitoses of a human meningioma were compared with chromosomal normal fibroblasts of the same patient. The meningioma cells showed an increased association frequency and a higher silver content of chromosome no. 14. These results also indicate that the loss of one or even more NORs can be compensated by a higher activity of the remaining NORs.

Variation in NOR Number

Since the early work of HEITZ (1931) it is well known that the secondary constrictions of chromosomes organize nucleolar material. However, the number of nucleolus organizers is not always constant in all individuals of the same species. For example viable mutants of *Xenopus laevis* are known which have only one instead of two NORs (MILLER & KNOWLAND 1970). The loss of two out of ten NORs is not very seldom observed in man bearing Robertsonian translocations (HURLEY & PATHAK 1977). In very few cases a Robertsonian translocation of even two pairs of acrocentrics was described in phenotypically normal men

(ORYE&DELIRE 1967, REEVES 1978: pers. commun.). A different number
of NORs in the same species is also known in plants (NICOLOFF&al. 1977).

This high variability of ribosomal cistrons arise the question, how the
constant supply with ribonucleoproteins is maintained in the cells. A
very interesting study concerning this question was performed by BERNS
& al. (1970): They destroyed one NOR in lung cells of a salamander by
argon laser beam. After this procedure they found in the daughter cells
two instead of three nucleoli. However these two nucleoli were obviously
enlarged and therefrom the authors concluded that the loss of one NOR
was compensated by an increased activity of the remaining NOR.

We had the opportunity to examine the effect of missing acrocentric
chromosomes on the function of NORs in human meningioma cells,
because, the cytogenetic study of a large series of meningiomas revealed
tumors with normal karyotype as well as tumors which had lost or
gained one, two or even more acrocentric chromosomes. We observed
that the number of nucleoli depended on the number of acrocentric
chromosomes whereas the nucleolar area remained constant (for details
see ZANKL & al. 1972, 1973).

In the last years further studies were performed in this field: MILLER
& al. (1977) reported that mutants of *Xenopus laevis* which possess only
one NOR produce the same amount of ribosomal RNA as normal
animals with two NORs. Cytological studies were especially carried out
in humans by examination of the satellite association phenomenon and
the use of selective NOR-staining procedures. The satellite associations,
that means the tendency of acrocentric chromosomes to stick together
with their satellite regions, are already known for some years and a
relationship was suggested to the nucleolar organization. A close
correlation was demonstrated by SCHMID & al. (1974) and HAYATA & al.
(1977) who reported that the frequency of satellite associations depends
on the extent of the NORs located in each acrocentric chromosome.
Furthermore it could be shown by *in situ* hybridization technique that
the association frequency is correlated with the number of ribosomal
cistrons (WARBURTON & al. 1976).

Silver Staining of NORs

A new way for investigations of the nucleolus organization was
opened by the development of staining techniques which allow selective
visualization of the nucleolus organizing regions. The first methods
which were based on extraction of DNA and histones brought
inconsistent results (MATSUI & SASAKI 1973, EIBERG 1974). HOWELL and
coworkers reported 1975 a simple silver staining method which is highly
reproducable: First some drops of "Ag-I-solution" (4 g $AgNO_3$ in 8 ml

aqua dest.) are given onto the chromosome preparation and the slide is mounted by a cover slip. By warming to 68 °C for some minutes the silver nitrate crystallizes around and beneath the coverglass which is afterwards removed. Two drops "Ag-II-solution" (4 g $AgNO_3$ dissolved into 5 ml aqua dest. and 7.5 ml NH_4OH) and of 3 % formaline are added to the air dried slide. The silver staining reaction is observed under phase microscope until the NORs become clearly visible and then the staining is stopped by rinsing of with destilled water. This method produces good results but has the disadvantage that the reaction is often to fast to stop

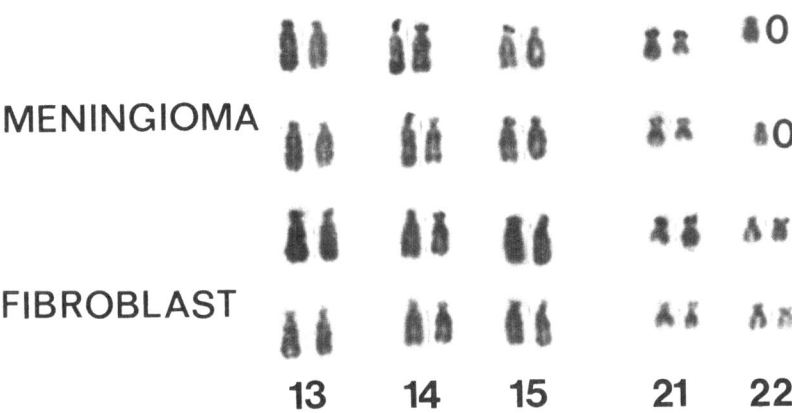

Fig. 1. NOR-staining and G-banding in the acrocentric chromosomes of mitoses from meningioma cells and fibroblasts

it at the optimal moment. Furthermore the staining must be observed under the microscope which is contaminated by silver nitrate. Therefore the modification of BLOOM & GOODPASTURE (1976) seems more practical to us. By this method the slides are only incubated overnight with 50 % silver nitrate under moist conditions at 37 °C.

Until now it is not exactly clarified which substances are stained by $AgNO_3$. Probably not the DNA of the ribosomal cistrons but the accumulated ribonucleiproteins react with the silver (SCHWARZACHER & al. 1978). Therefore inactive NORs should remain unstained. This was demonstrated by MILLER & al. (1976) who found that in most mouse-man hybrid cells all human NORs became inactivated and were not stained by silver nitrate. This inactivation also indicates functional relationships between the NORs of different species. Other studies showed a close correlation between the amount of silver at the NORs and the number of satellite associations (MILLER & al. 1977).

Original Results and Discussion

These reports stimulated us to examine how the loss of one or more NORs will influence the stainability of the remaining NORs and the association frequency. In our first study we used cells of a woman bearing a 13/13 Robertsonian translocation, which caused the loss of NORs of these chromosomes. The karyotype of these cells was compared

Table 1. Result of NOR-silver staining in mitoses with normal and 22-monosomic karyotypes (10 mitoses each)

45, XX, —22 (meningioma)	13^+	$13^{(+)}$	14^{++}	14^+	15^+	$15^{(+)}$	21^{++}	21^{++}	22^-	(22)
46, XX (fibroblast)	13^+	13^-	14^+	$14^{(+)}$	15^+	$15^{(+)}$	21^{++}	21^{++}	22^-	22^+

In all mitoses NOR unstained; $^{(+)}$ in some mitoses weak NOR-staining; $^+$ in most mitoses distinct NOR-staining; $^{++}$ in most mitoses strong NOR-staining.

Table 2. The number of associating chromosomes in mitoses with normal and 22-monosomic karyotypes (50 mitoses each)

Chromosome no.	21	22	13	14	15	total
45, XX, —22 (meningioma)	22	2^+	5	37^{++}	10	76
46, XX (fibroblast)	20	10	8	11	13	62

$^+$ Significant difference
$^{++}$ Highly significant difference } X^2-test

with that of her relatives who showed a normal chromosome set. It was observed that chromosome 22 showed a more intensive silver impregnation in the proband in comparison to her parents and her sister. The association frequency of this chromosome was also increased (for details see Zankl & Hahmann 1978).

22-monosomic meningioma cells seemed also to be a suitable material for such studies. Besides tumor cells we cultured also fibroblasts of the patient in order to have mitoses with normal karyotype. After chromosome preparation of both cultures the slides were silver stained and a Giemsa banding technique was performed

(ZANKL & BERNHARDT 1977). As shown in Table 1 and Fig. 1, chromosomes no. 14 accumulated more silver in its NO-region in 22-monosomic meningioma cells than in fibroblasts. These chromosomes showed also a higher association frequency (Table 2). The low number of associating chromosomes 22 in meningioma cells was probably caused by the loss of the one chromosome 22 which had a stainable NOR. These observations seem to indicate that the NORs of chromosomes 14 compensated the loss of the one NOR located in the missing chromosome 22. It must be stated however that the result might be influenced by the comparison of two different tissues, although MIKELSAAR & SCHWARZ-ACHER (1978) recently reported that the intensity of silver staining is not tissue-dependent. We hope to be able in the next future to examine a meningioma which contains cells with normal and 22-monosomic karyotypes.

This examination was supported by a grant of the *Deutsche Forschungsgemeinschaft*. Parts of this work are included in the doctoral thesis of H. HUWER.

References

BERNS, M. W., OHNUKI, Y., ROUNDS, D. E., OLSON, R. S., 1970: Modification of nucleolar expression following laser micro-irradiation of chromosomes. — Exp. Cell. Res. **60**, 133—138.

BLOOM, S. E., GOODPASTURE, C., 1976: An improved technique for selective silver staining of nuclear organizer regions in human chromosomes. — Hum. Genet. **34**, 199—206.

EIBERG, H., 1974: Satellite staining of human chromosomes. — Lancet **II**, 836.

HAYATA, I., OSHIMURA, M., SANDBERG, A. A., 1977: N-band polymorphism of human acrocentric chromosomes and its relevance to satellite association. — Hum. Genet. **36**, 55—61.

HEITZ, E., 1931: Die Ursache der gesetzmäßigen Zahl, Lage, Form und Größe pflanzlicher Nukleolen. — Planta **12**, 775—844.

HOWELL, W. M., DENTON, T. E., DIAMOND, J. R., 1975: Differential staining of satellite regions of human acrocentric chromosomes. — Experientia **31**, 260.

HURLEY, J. E., PATHAK, S., 1977: Elimination of nucleolus organizers in a case of 13/14 Robertsonian translocation. — Hum. Genet. **35**, 169—173.

MATSUI, S., SASAKI, M., 1973: Differential staining of nucleolus organizers in mammalian chromosomes. — Nature **246**, 148—150.

MIKELSAAR, A.-V., SCHWARZACHER, H. G., 1978: Comparison of silver staining of nucleolus organizer regions in human lymphocytes and fibroblasts. — Hum. Genet. **42**, 291—299.

MILLER, D. A., TANTRAVAHI, R., DEV, V. G., MILLER, O. J., 1977: Frequency of satellite association of human chromosomes is correlated with amount of Ag-staining of the nucleolus organizer region. — Amer. J. Hum. Genet. **29**, 490—502.

MILLER, L., KNOWLAND, J., 1970: Reduction of ribosomal RNA synthesis and ribosomal RNA genes in a mutant of *Xenopus laevis* which organizes only a partial nucleolus. — J. Molec. Biol. **53**, 321—328.

Miller, L., Daniel, J. C., 1977: Comparison of *in vivo* and *in vitro* ribosomal RNA synthesis in nucleolar mutants of *Xenopus laevis*. — In Vitro **13**, 557—563.

Miller, O. L., Miller, D. A., Dev, V. G., Tantravahi, R., Croce, C. M., 1976: Expression of human and suppression of mouse nucleous organizer activity in mouse-human somatic cell hybrids. — Proc. Nat. Acad. Sci. (U.S.) **73**, 4531—4535.

Nicoloff, H., Anastassova-Kristeva, M., Künzel, G., Rieger, R., 1977: The behavior of nucleolus organizers in structurally changed kartyotypes of barley. — Chromosoma **62**, 105—109.

Orye, E., Delire, C., 1967: Familial D/D and D/G translocation. — Helv. Paediatr. Acta **22**, 36—40.

Schmid, M., Krone, W., Vogel, W., 1974: On the relationship between the frequency of association and the nucleolar constriction of individual acrocentric chromosomes. — Hum. Genet. **23**, 267—277.

Schwarzacher, H. G., Mikelsaar, A.-V., Schnedl, W., 1978: The nature of the Ag-staining of nucleolus organizer regions. — Cytogen. Cell Genet. **20**, 24—39.

Warburton, D., Atwood, K. C., Henderson, A. S., 1976: Variation in the number of genes for rRNA among acrocentric chromosomes: correlation with frequency of satellite association. — Cytogen. Cell Genet. **17**, 221—230.

Zankl, H., Zang, K. D., 1972: The role of acrocentric chromosomes in nucleolar organization I. Correlation between the loss of acrocentric chromosomes and a decrease in the number of nucleoli in meningioma cell cultures. — Virchows Arch. Abt. B Zellpath. **11**, 251—256.

— — 1973: The role of acrocentric chromosomes in nucleolar organization II. Constancy of the nucleolar area independent of the number of acrocentric chromosomes in meningioma cell cultures. — Virchows Arch. Abt. B Zellpath. **13**, 113—118.

— Bernhardt, S., 1977: Combined silver staining of the nucleous organizing regions and Giemsa banding in human chromosomes. — Hum. Genet. **37**, 79—80.

— Hahmann, S., 1978: Cytogenetic examination of the NOR activity in a proband with 13/13 translocation and in her relatives. — Hum. Genet. **43**, 275—279.

Addresses of the authors: Prof. Dr. Heinrich Zankl, Department of Biology, The University of Kaiserslautern, D-6750 Kaiserslautern; Hanno Huwer, Institute of Human Genetics, The University of the Saarland, D-6650 Homburg/Saar, Federal Republic of Germany.

Chromatin Structure

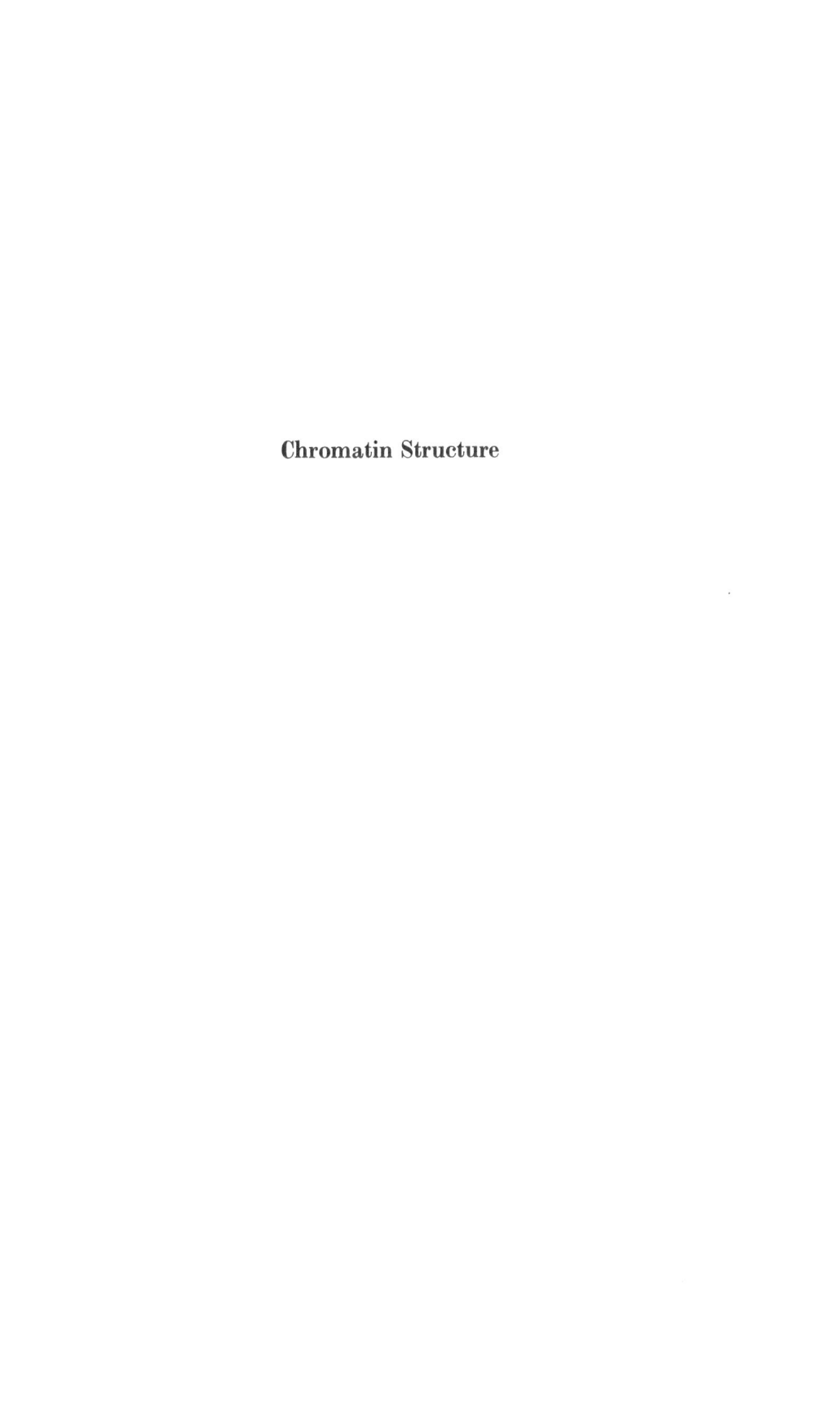

Pl. Syst. Evol., Suppl. 2, 187—199 (1979)

Institute of Biology, Department of Genetics,
The University of Tübingen, German Federal Republic

Nucleosomal Organization in Active and Inactive Plant Chromatin

By

Birgit Leber and **Vera Hemleben**, Tübingen

Key Words: Angiosperms, *Matthiola*, *Brassica*, *Phaseolus*, *Glycine*. — Plant chromatin, nucleosome, histone, ribosomal RNA genes, micrococcal nuclease, DNase I.

Abstract: Plant chromatin, like animal chromatin, can be cleaved by micrococcal nuclease digestion into nucleosomal subunits. The DNA content of the nucleosome is 175 $\pm$ 8 base pairs in all plant species and tissues studied. The DNA content of the core particle is 140 base pairs. Compared to animal histones, plant histone fractions H 1, H 2A and H 2B differ with respect to their electrophoretic mobility on polyacrylamide-SDS gels. H 2A and H 2B are species specific. H 1 is species and tissue specific. Studies using chromatin from tissues of different physiological activity do not indicate a different chromatin structure for active and inactive chromatin. Ribosomal genes in active and inactive state seem to be arranged in a configuration which can be cleaved by micrococcal nuclease attack into the same subunit pattern as the bulk of chromatin.

The basic organization of chromatin in the form of nucleoprotein subunits—called nucleosomes—now seems to be common to all eukaryotic organisms: A nucleosome consists of a core particle and a core connecting linker region. The nucleosome core is composed of two molecules of each of the histones H 2A, H 2B, H 3 and H 4 and approximately 140 base-pairs of DNA. The linker region, to which the histone H 1 is bound, can vary in length between 20 and 100 base-pairs (for a review see Thomas 1978). Compared to the increasing information on the internal nucleosomal structure and the higher order structure of animal chromatin, only little is known about plant chromatin (McGhee & Engel 1975, Gigot & al. 1976, Philipps & Gigot 1977).

Materials and Methods

Growth of the Plants. Seedlings of *Matthiola incana, Brassica pekinensis, Phaseolus aureus* and *Glycine max* were grown as described elsewhere (HEMLEBEN & al., 1975, GRIERSON 1974, GUILFOYLE & al. 1975). Flower petals of *Matthiola incana* (strain O 4 and O 6, Department of Genetics, Universität Tübingen) were harvested from plants with fully expanded flowers grown under field conditions.

Isolation of Nuclei. Nuclei were isolated from intact seedlings (*Matthiola incana* and *Brassica pekinensis*) or from roots, hypocotyls, cotyledons and primary leaves separately (*Phaseolus* and *Glycine*).

Plant tissue was homogenized with an Ultra-Turrax in the cold in a medium containing 0.4 M sucrose—0.05 M Tris—0.01 M MgCl$_2$ and 0.01 M NaCl, pH 7.8. The homogenate was filtered through miracloth. Triton X-100 was added to the filtrate (0.1 % final concentration) to lyse chloroplasts. The nuclei were pelleted by a centrifugation step (800 × g, 4 °C) for 5 min. The pellet was resuspended and washed in isolation buffer.

Nuclease Digestion. Nuclei were suspended in 500 µl of 10 mM Tris-HCl buffer, pH 7.5, containing 1 mM CaCl$_2$ and 250 mM sucrose. They were preincubated for 30 seconds at 37 °C and digested with micrococcal nuclease (Boehringer; 150 units/ml) for the indicated times. Nuclease treatment was terminated by adding 50 µl of 50 mM EDTA—50 mM EGTA, pH 7.5 and chilling on ice. After centrifugation at 3,000 × g for 5 min the first supernatant was discarded. The pellet of remaining nuclei was suspended in 1 ml of 1 mM EDTA, pH 7.5, lysed by brief homogenization and centrifuged at 3000 × g for 5 min. A second supernatant was recovered and analysed on isokinetic sucrose gradients.

For analysis of the DNA size distribution on agarose gels nuclease digestion was terminated by addition of 20 µl of 20 % SDS and the DNA was isolated by chloroform-isoamylalcohol extraction and ethanol precipitation.

Isokinetic Sucrose Gradients. 1 ml of the final digestion product (second supernatant) was layered on an isokinetic sucrose gradient ($c_t = 0.5$ M, $c_r = 1.7$ M and $V_m = 10$ ml; NOLL 1967) containing 1 mM EDTA, pH 7 and 100 mM NaCl. The gradients were centrifuged for 15 hours at 36,000 rpm and 4 °C in a Beckman SW 40 rotor. 6-drop fractions were collected from the top using an Isco density fractionator, model 640. For subsequent studies fractions were pooled and treated with ribonuclease (20 µg/ml) for 30 min. The DNA was isolated by chloroform-isoamylalcohol extraction and ethanol precipitation.

To identify newly synthesized RNA attached to the chromatin *Matthiola* seedlings were grown in [5-^{3}H] uridine (26 Ci/mmole; 150 µCi/ml) for 20 hours. Nuclei were isolated and subjected to the procedures described above.

Analysis of DNA Fragments on Agarose Gels. DNA fragments generated by micrococcal endonuclease digestion were separated by slab-gel electrophoresis on 2.5 % agarose gels. The running buffer was 0.04 M Tris—0.02 M sodium acetate—1 mM EDTA, pH 7.8. Electrophoresis was for 6 hours at 140 V.

Gels were stained for 30 min with 1 µg/ml ethidium bromide dissolved in running buffer and photographed under UV illumination. Molecular weight markers were DNA fragments of Hae III digested λdv 1 DNA (kindly supplied by Dr. R. STREECK).

Blotting and Hybridization With ^{125}I-RNA. DNA fragments were transferred from the agarose gels to nitrocellulose filters according to the method of SOUTHERN (1975). The rRNA used for subsequent hybridization studies was iodinated *in vitro* (GETZ & al. 1972, OROZ & WETMUR 1974). Hybridization was performed at 60 °C for 16-18 hours with [125]I-18 S rRNA + 25 S rRNA (0.2 µg/ml; specific radioactivity 4.5 × 10[6] cpm/µg) in 6 × SSC containing 25 % formamide. After the incubation filters were washed three times with 2 × SSC (10 min), treated with ribonuclease (20 µg/ml; 45 min) washed again three times in 2 × SSC (10 min) and dried. Filters were exposed on a Kodak X-ray film (No Screen Film) for two weeks.

Actinomycin-CsCl Gradients. Total DNA of *Matthiola incana* isolated as described by HEMLEBEN & al. (1975) was centrifuged in an actinomycin-caesium chloride gradient according to HEMLEBEN & al. (1977) in order to separate ribosomal DNA sequences from the bulk of DNA.

DNA-RNA Hybridization. DNA isolated from isokinetic sucrose gradient fractions dissolved in 1 ml 1 × SSC or fractions from actinomycin-CsCl gradients diluted with 1 ml gradient buffer were denatured, neutralized, and loaded onto nitrocellulose filters according to HEMLEBEN & al. (1977). Hybridization was carried out in 6 × SSC at 60 °C for 4 hours using either [3]H-25 S and 18 S rRNA (specific radioactivity 40,000 cpm/µg; 2 µg/ml) or ^{3}H uridine labelled RNA isolated from digested chromatin (specific radioactivity 700 cpm/µg; 20 µg/ml). The following washing procedures, RNase-treatment and DNA determination were performed as described previously (SPRING & al. 1978).

Radioactively labelled rRNA samples used for hybridization experiments were obtained by growing *Matthiola* seedlings in [5-^{3}H] uridine (500 µCi/ml) for 3 days. RNA was extracted and 25 S and 18 S rRNA were purified by polyacrylamide gel electrophoresis (GRIERSON & HEMLEBEN 1977).

Extraction of Histones. Isolated nuclei were suspended by homogenization in 0.25 N HCl and stirred for 2 hours in the cold. Non-histones were pelleted by a centrifugation step (10,000 × g, 30 min, 10 °C). From the supernatant histones were precipitated with 9 volumes of acetone at —27 °C for 48 hours. After centrifugation (10,000 × g, 30 min, 10 °C) the histone pellet was washed extensively with ethanol and air dried.

Polyacrylamide Gel Electrophoresis. Electrophoretic separation of plant histones was carried out on polyacrylamide sodium dodecyl sulfate slab gels (LAEMMLI 1970). The separating gel contained 15 % acrylamide, the stacking gel 3 % acrylamide. Electrophoresis was at room temperature at constant voltage (10 V/cm) with bromphenol blue as tracking dye. Gels were stained for 1 hour at 60 °C with 0.25 % Coomassie Brilliant Blue R in 50 % methanol—10 % acetic acid and destained in 30 % methanol—10 % acetic acid. Calf thymus histones (Boehringer) were used as molecular weight markers.

Results and Discussion

Much of our present knowledge concerning the organization of eukaryotic chromatin has come from digestion experiments using the Ca[++]-dependent nuclease from *Staphylococcus aureus:* Limited digestion of plant nuclei with micrococcal nuclease results in a mixture of

Birgit Leber and Vera Hemleben:

monomeric and multimeric nucleosome particles, i.e. the micrococcal nuclease preferentially cuts whithin the linker region. The generated subunits can be separated by centrifugation in isokinetic sucrose gradients (Fig. 1). The separation is improved by gradients containing a

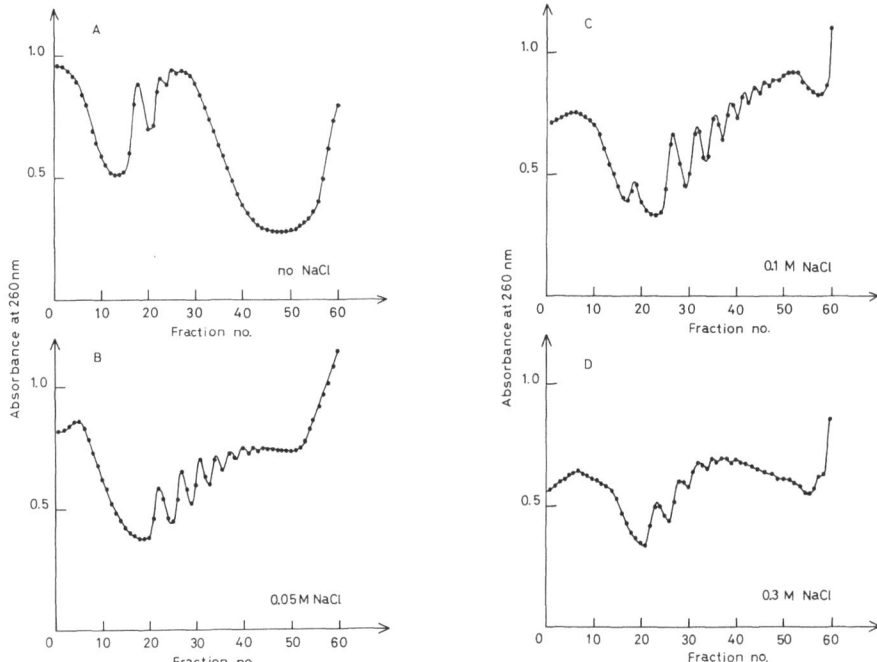

Fig. 1. Effect of salt on the fractionation of chromatin subunits. Nuclei from *Matthiola* flower petals were digested for 4 min with 150 units/ml of micrococcal endonuclease. The second supernatant (see Materials and Methods) was layered onto an isokinetic sucrose gradient. The gradient buffer contained A) no NaCl, B) 0.05 M NaCl, C) 0.1 M NaCl, D) 0.3 M NaCl

salt concentration between 0.05 M and 0.10 M NaCl (Figs. 1 *B* and 1 *C*) as compared to gradients containing no NaCl (Fig. 1 *A*) or 0.30 M NaCl (Fig. 1 *D*). A similar result was obtained for Hela cells (Whitlock & Simpson 1976). After deproteinisation of the monomeric and multimeric fractions, size analysis of the DNA fragments on a 2.5 % agarose gel results in DNA with increasing fragment lengths following the direction of sedimentation (Fig. 2).

Electrophoresis of DNA from unfractionated nuclease-treated plant chromatin on a 2.5% agarose gel produces a characteristic repeat pattern (Fig. 3). In addition to whole seedlings of *Matthiola* and *Brassica*, and flower petals of *Matthiola*, we studied the chromatin of roots, hypocotyls, cotyledons and primary leaves of *Glycine* and *Phaseolus* seedlings in detail. Differences in the repeat length were

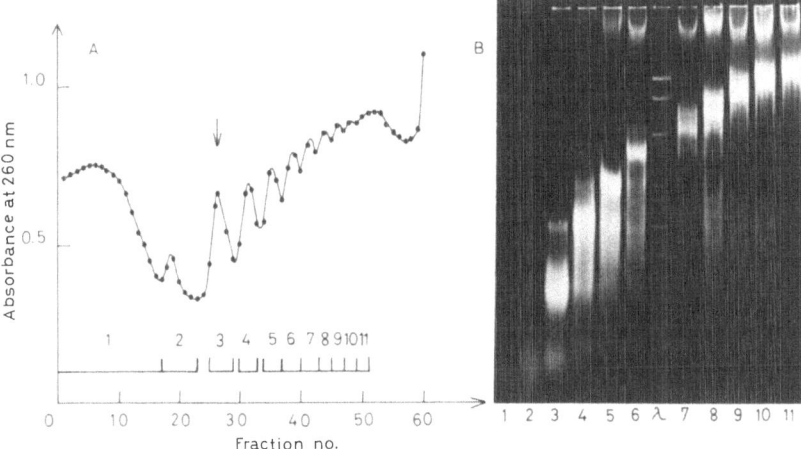

Fig. 2.   Size distribution of DNA fragments derived from fractionated chromatin subunits. Nuclease digested *Matthiola* chromatin (4 min digest) was fractionated on an isokinetic sucrose gradient (A). DNA was isolated from the pooled fractions (see inserts) and separated on a 2.5% agarose gel (B) together with Bsp-digested λdv1 DNA. (Arrow shows position of the nucleosome monomer)

neither observed among different plant species nor between different tissues of the same plant. The molecular weight of the DNA from the monomer is 175 ± 8 base-pairs as calibrated by Bsp-cleaved bacteriophage λdv 1 DNA. This value is close to that described for *Neurospora* (NOLL 1976) and *Physarum* (JOHNSON&al. 1976), but lower than the data reported for tobacco and barley (PHILIPPS & GIGOT 1977). The core particle contains 140 base-pairs of DNA as observed for all protozoans, fungi and higher eukaryotic organisms, which were studied so far.

Concerning the five histone classes H 1 (very lysine-rich), H 2A and H 2B (lysine-rich), H 3 and H 4 (arginine-rich), only the last two plant histones are nearly identical to those of calf thymus confirming their high evolutionary conservativism (Fig. 4). The H 1 fraction of plants shows a slower migration on polyacrylamide-SDS gels, and a more

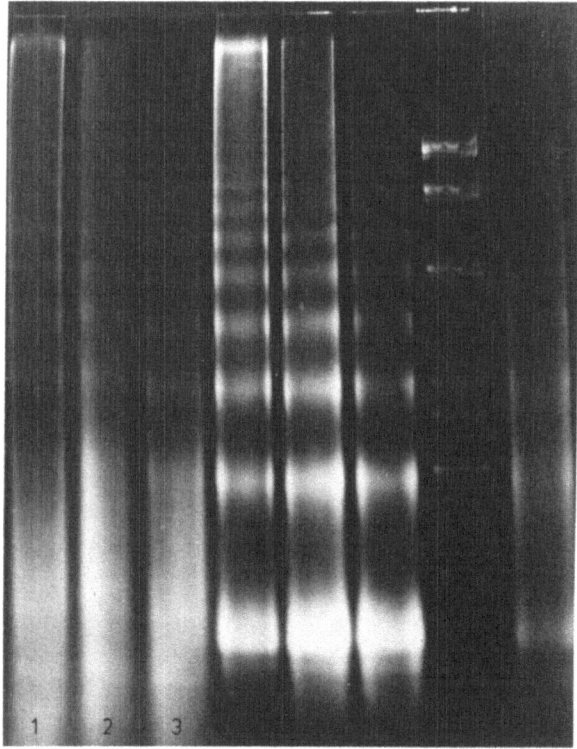

Fig. 3. Nucleosomal DNA pattern of chromatin from *Matthiola* seedlings (slots 1-3), *Matthiola* flower petals (slots 4-6) and *Brassica* seedlings (slot 8) as obtained by electrophoresis on 2.5% agarose gels. Micrococcal digestion was for 3 min (slots 1 and 4). 6 min (slots 2 and 5), 10 min (slots 3 and 6) and 5 min (slot 8). Molecular weight markers were DNA fragments of 1686, 1320, 881, 535, 462, 357, 271, 230, 215, 180, 144, 133, 84, 38 and 34 base-pairs of Hae III-digested λ dv1 DNA

complex electrophoretic pattern than the corresponding animal histone. No correspondence can be found between the faster migrating plant histones now called PH 1 and PH 2 (Nadeau & al. 1974) and the animal histone fractions H 2A and H 2B. They appear to be species-specific. H 1 is species- and tissue-specific (data not shown).

The complex of DNA and histones supporting the compact structure of the nucleosome (packing ratio 5-7) is the first level of packing DNA in the nucleus. Further interest was focussed on the nucleosome structure with respect to eukaryotic gene regulation: Do genes, which are active in

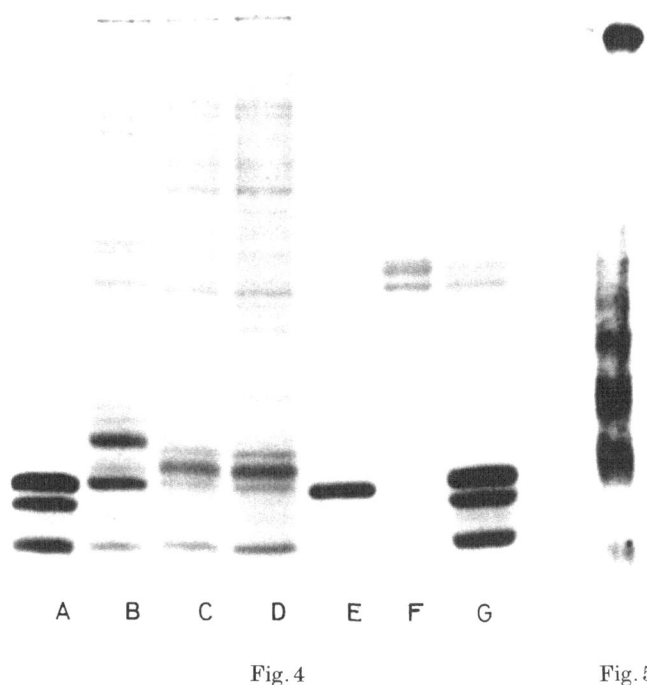

Fig. 4   Fig. 5

Fig. 4. Electrophoretic mobility of plant histones compared to calf thymus histones. Plant histones were isolated from nuclei of *Brassica* and *Glycine*, and run on a 15% polyacrylamide-SDS gel for 6 hours at 140 V. A) total histones from calf thymus, B) total histones from *Brassica*, C) and D) total histones from *Glycine*, E) histone H 2B, F) histone H 1 and G) total histones from calf thymus

Fig. 5. Distribution of ribosomal DNA sequences in nuclease digested plant chromatin. Nuclei of *Brassica pekinensis* were digested for 7 min with micrococcal nuclease (150 units/ml). DNA fragments were isolated, separated on 2.5% agarose slab gels and transferred to nitrocellulose filters according to the technique of SOUTHERN. Ribosomal DNA sequences were detected by hybridization using ¹²⁵I-labelled 18 S and 25 S rRNA followed by autoradiography

the process of transcription, have a nucleosomal structure, and if they do, are active nucleosomes identical to those of bulk chromatin?

Our studies on DNA repeat length and histone composition of plant chromatin from various sources have shown that there are no significant differences between tissues of more, and tissues of less, physiological activity. A more direct approach to identify active genes was performed

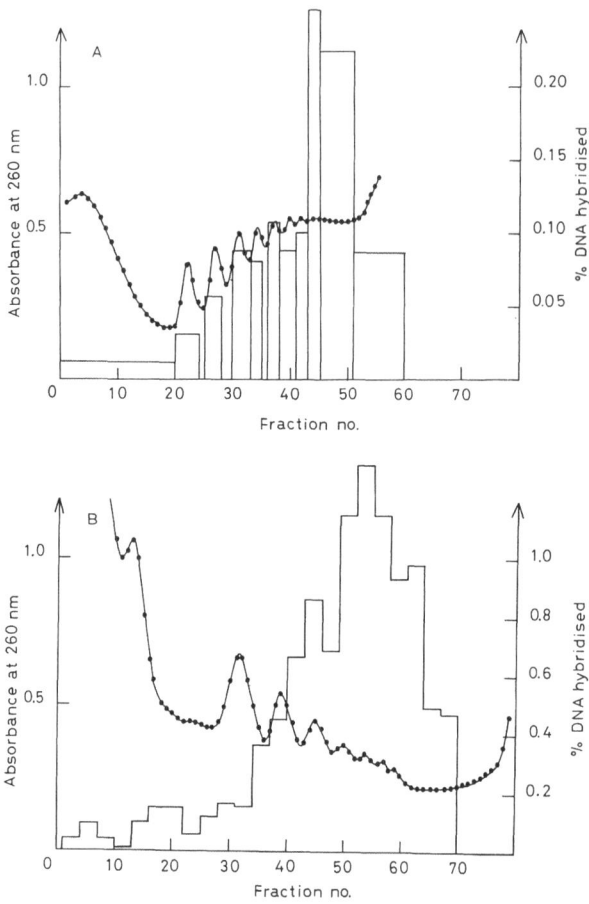

Fig. 6. Hybridization of ³H-rRNA to DNA from chromatin subunits fractionated on an isokinetic sucrose gradient. Nuclei were isolated from A) *Matthiola* flower petals and B) *Brassica* seedlings, and digested with micrococcal nuclease (4 min digestion 150 units/ml). Chromatin subunits were fractionated on an isokinetic sucrose gradient containing 0.1 M NaCl. The DNA of the pooled peak fractions was isolated, denatured and fixed onto nitrocellulose filters. Filters were incubated for 4 hours at 60 °C in 6 × SSC containing 2 μg/ml 25 S and 18 S rRNA. ●—●—●—●—●—●—●—   Absorbance   at   260 nm; ——————————— % DNA hybridized

studying the structure of nucleolar chromatin compared to nuclear chromatin. Ribosomal genes of higher plants have two advantages in comparison to unique sequences. They are very numerous and of high transcriptional activity in growing and differentiating tissues.

Nuclei isolated from *Brassica* seedlings were digested for 7 min with micrococcal nuclease and the generated DNA fragments were fractionated on 2.5% agarose gels. The DNA was transferred onto nitrocellulose filters following the method of SOUTHERN (1975) and hybridized to ^{125}I-labelled rRNA (Fig. 5). The autoradiogram shows a distinct repeat pattern coinciding exactly with the electrophoretic pattern of the ethidium bromide stained DNA fragments. Therefore, ribosomal genes appear to be present within the nucleosomal fraction of nuclease digested chromatin.

Further evidence was obtained by the following experiment additionally providing some quantitative information. Nuclease-digested chromatin from *Matthiola* flower petals was fractionated on an isokinetic sucrose gradient and the peak fractions were pooled. DNA was isolated, denatured, fixed onto nitrocellulose filters and hybridized to ^{3}H-labelled 18S and 25S ribosomal RNA in excess. Hybridization— expressed as percentage of DNA hybridized to rRNA—is observed along the whole sedimentation profile (Fig. 6 *A*). Ribosomal DNA sequences, however, seem to be concentrated in the region of subunit multimers comprising 7 and more nucleosomes. Ribosomal genes, therefore, appear to be less sensitive to endonuclease attack than the bulk of chromatin, possibly because of a modified nucleosomal structure.

Principally the same result is obtained if a comparable experiment is performed with *Brassica* seedlings (Fig. 6 *B*). Chromatin of seedlings is thought to be of higher transcriptional activity than chromatin of fully differentiated *Matthiola* flower petals. Despite the more proceeded nuclease digestion, again the highest hybridization rate is found in the region of nucleosomal multimers. But prolonged incubation with micrococcal nuclease results in a shift of the hybridizing fractions towards the monomeric region (data not shown).

So far no evidence has been presented for the transcriptionally active state of the ribosomal genes studied. A gene which is active in transcription consists of a DNA matrix, the associated RNA polymerase, and the attached nascent ribonucleoprotein fibril. To identify such a transcriptional complex, *Matthiola* seedlings were labelled with [3H] uridine *in vivo*. Nuclei of these seedlings were isolated and treated with micrococcal nuclease. The digested chromatin was fractionated on an isokinetic sucrose gradient and the radioactivity of each fraction was measured (Fig. 7 *A*).

Radioactively labelled, newly synthesized RNA is found in all fractions of the repeat pattern. To exclude artifacts and to show whether the RNA is actually bound to the matrix DNA, the RNA from the digested chromatin region (fractions 20-40, Fig. 7 *A*) was isolated and recentrifuged on a comparable gradient. In order to identify unspecific

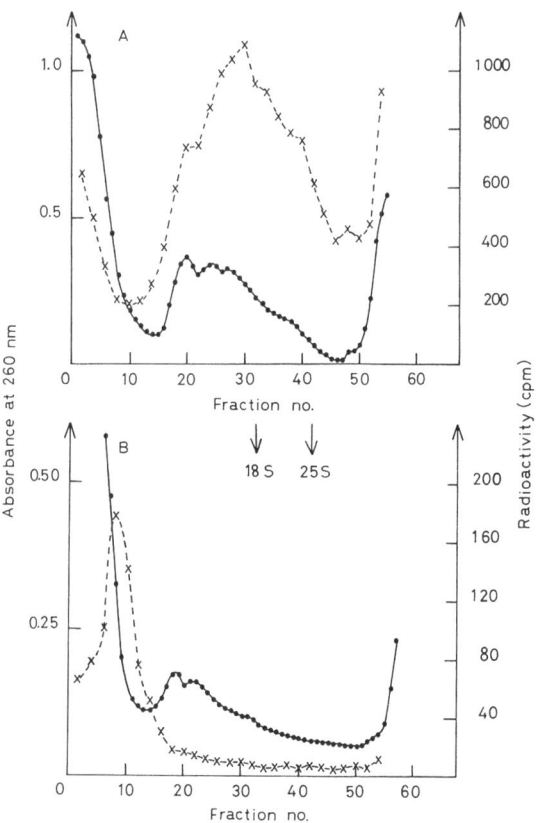

Fig. 7. Detection of newly transcribed RNA within fractionated chromatin subunits. *Matthiola* seedlings were incubated for 20 hours with [³H]uridine *in vivo*. Chromatin derived from nuclease digested nuclei of these seedlings was fractionated on an isokinetic sucrose gradient. RNA sequences were identified by their radioactive label (A). The size of the ³H-labelled RNA attached to chromatin subunits was determined by recentrifugation on an isokinetic sucrose gradient (B). ●—●—●—●—●—●—●— Absorbance at 260 nm; ×——×——×——× ³H-radioactivity

aggregation between ³H-RNA and chromatin, unlabelled digested chromatin was centrifuged in the same gradient (Fig. 7 *B*). ³H-labelled RNA appears exclusively in the region of low molecular weight corresponding to about 5 S. This small size indicates either degradation during nuclease digestion, or unfinished transcripts. Clearly no unspecific attachment to the added chromatin occurred. The newly synthesized RNA derived from the nucleosomal fractions of the isokinetic gradient (Fig. 7 A) must have been originally bound to the DNA matrix. To see whether ribosomal RNA sequences are present in this RNA preparation, a further experiment was performed. Total plant DNA was centrifuged in a CsCl density gradient containing actinomycin (Fig. 8). Actinomycin binds specifically to GC-rich regions which are reduced in their density and can be separated from the bulk of DNA. GC-

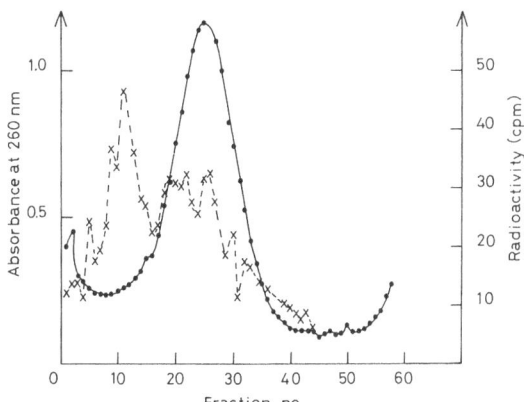

Fig.˙ 8. Identification of ribosomal sequences of the newly synthesized ³H-labelled RNA attached to fractionated chromatin. Unlabelled total DNA from *Matthiola incana* was fractionated on an actinomycin-CsCl gradient. Each fraction was loaded onto filters and hybridized to ³H-labelled RNA (20 µg/ml, specific radioactivity: 700 cpm/µg; 16 hours, 60 °C, in 6 × SSC) isolated from chromatin subunits (fractions 20-40) of the sucrose gradient shown in Fig.7.
●—●—● Absorbance at 260 nm × — — × — — × ³H-radioactivity

rich ribosomal sequences are located on the light side of the mainband (HEMLEBEN&al. 1977). The DNA of every single fraction was denatured, fixed on nitrocellulose filters and hybridized to the above characterized ³H-labelled RNA (Fig. 8). The diagram shows that part of the radioactive RNA hybridizes to those fractions of the gradient, which contain the ribosomal DNA sequences. The remaining RNA of nonribosomal origin hybridizes to main-band DNA.

So far studied ribosomal genes in active and inactive state clearly appear to be arranged in a configuration which can be cleaved by nuclease attack into the same subunit pattern as the bulk of chromatin. Despite contradictions coming from electron microscopic studies (FRANKE&al. 1976; LAIRD&al. 1976), these observations appear to be in agreement with results obtained for ribosomal genes of *Tetrahymena*

(MATHIS&GOROVSKY 1976, GRAINGER&OGLE 1978) and *Xenopus* (REEVES 1976), and for unique sequences of animal cells (for a review see LILLEY 1978). Since the response of transcriptionally active sequences to digestion by DNase I seems to be strikingly different from that of inactive chromatin (WEINTRAUB&GROUDINE 1976, GAREL&AXEL 1976, STALDER&al. 1978), DNase I digestion of plant nuclei may be a suitable approach to characterize active genes.

We are grateful to Miss ANNE ROTH for excellent technical assistance.—The work has been supported by the *Deutsche Forschungsgemeinschaft*.

References

FRANKE, W. W., SCHEER, U., TRENDELENBURG, M. F., SPRING, H., ZENTGRAF, H., 1976: Absence of nucleosomes in transcriptionally active chromatin. — Cytobiologie **13**, 401—434.

GAREL, A., AXEL, R., 1976: Selective digestion of transcriptionally active ovalbumine genes from oviduct nuclei. — Proc. Nat. Acad. Sci. (U.S.) **73**, 3966—3970.

GETZ, M. J., ALTENBURGER, L. C., SAUNDERS, G. F., 1972: The use of RNA labelled *in vitro* with iodine-125 in molecular hybridization experiments. — Biochim. Biophys. Acta **287**, 485—494.

GIGOT, C., PHILIPPS, G., NICOLAIEFF, A., HIRTH, L., 1976: Some properties of tobacco protoplast chromatin. — Nuc. Acids Res. **3**, 2315—2329.

GRAINGER, R. M., OGLE, R. C., 1978: Chromatin structure of the ribosomal RNA genes in *Physarum polycephalum*. — Chromosoma **65**, 115—515.

GRIERSON, D., 1974: Characterization of ribonucleic acid components from leaves of *Phaseolus aureus*. — Europ. J. Biochem. **44**, 509—515.

— HEMLEBEN, V., 1977: Ribonucleic acid from the higher plant *Matthiola incana*. Molecular weight measurements and DNA·RNA hybridization studies. — Biochim. Biophys. Acta **475**, 424—436.

GUILFOYLE, T. J., LIN, C. Y., CHEN, Y. M., NAGAO, R. T., KEY, J. L., 1975: Enhancement of soybean polymerase I by auxin. — Proc. Nat. Acad. Sci. (U.S.) **72**, 66—72.

HEMLEBEN, V., ERMISCH, N., KIMMICH, D., LEBER, B., PETER, G., 1975: Studies on the fate of homologous DNA applied to seedlings of *Matthiola incana*. — Eur. J. Biochem. **56**, 403—411.

— GRIERSON, D., DERTMANN, H., 1977: The use of equilibrium centrifugation in actinomycin caesium chloride for the purification of ribosomal DNA. — Pl. Sci. Lett. **9**, 129—135.

JOHNSON, E. M., LITTAU, V. C., ALLFREY, V. G., BRADBURY, E. M., MATTHEWS, H. R., 1976: The subunit structure of chromatin from *Physarum polycephalum*. — Nucl. Acids Res. **3**, 3313—3329.

LAEMMLI, U. K., 1970: Cleavage of structural proteins during the assembly of the head of bacteriophage T4. Nature **227**, 680—685.

LAIRD, C. D., WILKINSON, L. E., FOE, V. E., CHOOI, W. Y., 1976: Analysis of chromatin-associated fiber arrays. — Chromosoma **58**, 169—192.

LILLEY, D. M. J., 1978: Active chromatine structure. — Cell Biol. Intern. Rep. **2**, 1—10.

MATHIS, D. J., GOROVSKY, M. A., 1976: Subunit structure of rDNA-containing chromatin. — Biochemistry 15, 750—755.
McGHEE, J. D., ENGEL, J. D., 1975: Subunit structure of chromatin is the same in plants and animals. — Nature 254, 449—450.
NADEAU, P., PALOTTA, D., LAFONTAINE, J. G., 1974: Electrophoretic study of plant histones: Comparison with vertebrate histones. — Arch. Biochem. Biophys. 161, 171—177.
NOLL, H., 1967: Characterization of macromolecules by constant velocity sedimentation. — Nature 215, 360—363.
NOLL, M., 1976: Differences and similarities in chromatin structure of Neurospora crassa and higher eukaryotes. — Cell 8, 349—355.
OROZ, I. M., WERTMUR, J. B., 1974: In vitro iodination of DNA. Maximizing iodination while minimizing degradation; use of buoyant density shifts for DNA-DNA hybrid isolation. — Biochemistry 13, 5467—5473.
PHILIPPS, G., GIGOT, C., 1977: DNA associated with nucleosomes in plants. — Nucl. Acids Res. 4, 3617—3626.
REEVES, R., 1976: Ribosomal genes of Xenopus laevis: Evidence of nucleosomes in transcriptionally active chromatin. — Science 194, 529—532.
SOUTHERN, E. M., 1975: Detection of specific sequences among DNA fragments separated by gel-electrophoresis. — J. Molec. Biol. 98, 503—515.
SPRING, H., GRIERSON, D., HEMLEBEN, V., STÖHR, M., KROHNE, G., STADLER, J., FRANKE, W. W., 1978: DNA contents and numbers of nucleoli and pre-rRNA-gens in nuclei of gametes and vegetative cells of Acetabularia mediterranea. — Exp. Cell Res. 114, 203—215.
STADLER, J., SEEBECK, T., BRAUN, R., 1978: Degradation of the ribosomal genes by DNase I in Physarum polycephalum. — Eur. J. Biochem. 90, 391—395.
THOMAS, J. O., 1978: Chromatin structure. — In CLARK, B. F. C., (Ed.): International Review of Biochemistry, Biochemistry of Nucleic Acids II, Vol. 17, 181—232. — Baltimore: University Park Press.
WEINTRAUB, H., GROUDINE, M., 1976: Chromosomal subunits in active genes have an altered conformation. — Science 193, 848—856.
WHITLOCK, J. P., SIMPSON, R. T., 1976: Preparation and physical characterization of a homogenous population of monomeric nucleosomes from HeLa cells. — Nucl. Acids Res. 3, 2255—2266.

Address of the authors: Dr. BIRGIT LEBER and Prof. Dr. VERA HEMLEBEN, Institut für Biologie II, Lehrstuhl für Genetik der Universität Tübingen, Auf der Morgenstelle 28, D-7400 Tübingen, Federal Republic of Germany.

Pl. Syst. Evol., Suppl. 2, 201—219 (1979)
© by Springer-Verlag 1979

Institute of Physiological Chemistry,
Philipps University, Marburg/Lahn, Federal Republic of Germany

Analysis of Accessible Sites in Modified Nucleosomes

By

Detlef Doenecke, Marburg/Lahn

Key Words: Chromatin and nucleosomes (from Wistar rats), micrococcal nuclease, modification methylase (from *Haemophilus influenzae*), polylysine, ethidium bromide, formaldehyde, dimethylsuberimidate, polycrylamide gel electrophoresis.

Abstract: Methylases from the bacterium *Haemophilus influenzae* are proposed as a tool for a selective modification of the internucleosomal linker DNA. The exclusive methylation at this site was demonstrated by a comparison of the rate of degradation of radioactively methylated chromatin and of total DNA with micrococcal nuclease. Analysis of the DNA fragment pattern after digestion of methylated nucleosome dimers reveals that only full size nucleosome monomers and larger fragments (e.g. dimers) are radioactively labelled, whereas smaller (subnucleosomal) DNA cleavage products do not carry any methylated portions any more. Methylation and digestion studies in the presence of polylysine support the conclusion that modification methylases and micrococcal nuclease act at the internucleosomal linker DNA, whereas polylysine interacts preferentially with the nucleosomal core DNA.—The minor groove of the nucleosomal core DNA was identified as the site of polylysine binding by competition studies with ethidium bromide, since intercalation of this molecule proceeds via the minor groove. Scatchard plots of ethidium binding parameters and analysis of nucleosome cleavage patterns indicate a competition of polylysine and ethidium bromide for binding sites in the nucleosomal core DNA.—Finally, cross-linking studies with formaldehyde and dimethylsuberimidate suggest that the conformational change of the nucleosomal DNA due to ethidium intercalation leads to slight changes in the mode of interaction of the different nucleosomal compounds.

Electron microscopy, biochemical and biophysical studies have established in the years since 1973 that chromatin is organized as a regular chain of repeated subunits, termed nucleosomes (OUDET & al. 1975) or ν-bodies (OLINS & OLINS 1974). These consist of flat, somewhat wedge-shaped histone/DNA complexes (FINCH & al. 1977), the nucleosome cores, and they are connected by a linker DNA segment. This latter part of the nucleosomal DNA is associated with histone H 1, whereas two of each of the other four histones form an octamer and are bound to the 140 nucleotide pairs of the nucleosome core (for review, see KORNBERG 1977).

Nucleosomes have been detected, isolated and analysed as to their internal structure by means of several nucleases. An endogenous nuclear enzyme (HEWISH & BURGOYNE 1973), micrococcal nuclease (NOLL1974a) and streptodornase (NOLL 1977) preferentially cleave internucleosomally, but they also split the nucleosomal core DNA at defined sites. DNase I (NOLL 1974b), DNase II and micrococcal nuclease (SOLLNER-WEBB & al. 1978) cleave at staggered sites on each strand of the nucleosomal core and the linker DNA double helix at 10 nucleotide intervals. Under certain conditions, DNase II cleaves the nucleosomal DNA internally in a way, which yields DNA fragments of half-nucleosomal size (ALTENBURGER & al. 1976). In addition, this enzyme cleaves preferentially in such regions of chromatin, which are activated for RNA synthesis (GOTTESFELD 1978). These chromatin portions, where RNA synthesis occurs, do not look like chains of nucleosomes in the electron microscope. Micrococcal nuclease (and DNase II, SOLLNER-WEBB & al. 1978) cleaves however at regular intervals of nucleosomal size and the resulting particles, which carry activated genes, are very sensitive to DNase I (WEINTRAUB & GROUDINE 1976).

Thus, transcriptionally active chromatin also seems to show repeated segments (as revealed by nuclease studies, GAREL & AXEL 1976, WEINTRAUB & GROUDINE 1976), but they are structurally different compared with nucleosomes in transcriptionally inactive portions of the genome (FRANKE & SCHEER 1978). The elucidation of the structural basis for the differences in transcriptionally active and inactive chromatin segments requires further methods for the analysis of chromatin subunits.

The structure of a chromatin segment depends on the mode of interaction of proteins and DNA. This is in part reflected in the accessibility of both the protein and the DNA moiety of chromatin. Nucleases (see above) and proteases (e.g. SAHASRABUDDHE & VAN HOLDE 1974, WEINTRAUB 1975, KATO & IWAI 1977, DOENECKE 1978) have been succesfully used for this purpose, but the hydrolysis of the phosphodiester and peptide bonds in DNA and proteins respectively, is inevitably accompanied by a destruction of the accessible site.

This is the reason, why we want to propose the use of additional tools for the analysis of accessible sites in chromatin. Modification of bases in DNA, limited protection against cleavage and induction of alterations in cleavage patterns will be proposed as additional tools in the analysis of the structure of nucleosomes.

Materials and Methods

Materials. Rat livers were taken from male Wistar rats. *Haemophilus influenzae* RD. was obtained from MRE, Porton Down, Bucks. (U.K.). All reagents for electrophoresis (acrylamide, N,N′-methylene bisacrylamide, N,N,N′,N′-tetramethyl ethylenediamine, sodium dodecylsulfate, "stains all" and amido black) were from Serva, Heidelberg. Poly-L-lysine hydrobromide (M_r 3,400) and dimethylsuberimidate were purchased from Sigma, München, nuclease-free sucrose and ethidium bromide from Serva, Heidelberg. S-adenosyl-L-(methyl-^{3}H)methionine (7.5 Ci/mmol) was obtained from Amersham-Buchler (Braunschweig). The enzymes used were either from Boehringer (Mannheim) or Worthington (Freehold, N.J.) with similar results. All other reagents were (reagent grade) from Merck A.G. (Darmstadt).

Preparation and Fractionation of Chromatin. Rat liver nuclei were prepared by a modified [DOENECKE (1975)-CHAUVEAU (1956)] procedure and suspended in the digestion medium A first described by HEWISH & BURGOYNE (1973). It contains spermine and spermidine but no divalent cations and thus prevents endogenous enzymes from cleaving chromatin DNA. The nuclei in buffer A were made 1 mM in $CaCl_2$ and incubated with micrococcal nuclease (100 units/ml) for 5 up to 10 minutes at 37 °C (NOLL & al. 1975). The incubation was stopped with Na_2EDTA (2 mM) and the subsequent steps (lysis of the nuclei and fractionation of chromatin on isokinetic (NOLL 1967) sucrose gradients) was done as described by FINCH & al. (1975). The individual peaks of nucleosome monomers, dimers and trimers were dialyzed against buffer B (5 mM sodium phosphate buffer, pH 6.8, 0.2 mM Na_2EDTA), concentrated in Minicon B 15 cells (Amicon, Witten) and recentrifuged.

Fixation of Nucleosomes With Formaldehyde. Nucleosomes or modified nucleosomes (see below) in buffer B were dialyzed against buffer B supplemented with 1 % formaldehyde for 24 hours at + 4 °C (BRUTLAG & al. 1969, DOENECKE 1978), then dialysis was continued for 48 hours against buffer B (six changes) and then the crosslinked material was subjected to some of the procedures described in the other paragraphs.

Cross-Linking of the Nucleosomal Core Histones. The nucleosomal core histones were cross-linked with dimethylsuberimidate (DAVIES & STARK 1970) as described by KORNBERG & THOMAS (1974) and modified by STEIN & al. (1977), with little modifications. Nucleosome cores, prepared as described previously (CLIMENT & al. 1977), in 0.01 M borate buffer (pH 9.4) were treated with dimethylsuberimidate (final concentration 10 mg/ml) for 20 minutes at room temperature. During the addition of the reagent to the nucleosome core preparation (A 260 = 2.0), the pH was constantly monitored at 9.4 by dropwise addition of 0.1 M NaOH. At the end of the incubation, the sample was chilled and put on isokinetic sucrose gradients (5 to 28.8 %) in buffer C (25 mM Tris-HCl, pH 8.0, 1 mM Na_2EDTA). After 22 hours at 90,000 × g in a Beckman SW 27 rotor, the peak fractions were collected and dialyzed against buffer B.

Cesium Chloride Gradient Centrifugation. Cesium chloride gradients of formaldehyde-fixed chromatin or chromatin/polylysine complexes were performed in the Beckman SW 60 rotor as described (DOENECKE & McCARTHY 1976, 1977, DOENECKE 1976 a, 1978).

Preparation of Modification Methylases. Modification methylases from *Haemophilus influenzae* RD. were prepared as described by ROY & SMITH (1972)

with slight modifications (DOENECKE & MCCARTHY 1977). The different peak
fractions after the first phosphocellulose chromatography step were pooled,
precipitated with ammonium sulfate at 50 % saturation and dissolved in buffer
D (10 mM potassium phosphate, pH 7.4, 5 % glycerol, 5 mM 2-mercaptoethanol,
0.5 mM Na$_2$EDTA), made 50 % (v/v) in glycerol and stored in liquid nitrogen
until use.

Methylation of Free and Nucleosomal DNA. Protein-free DNA or nucleosomes
were methylated with the modification enzymes from *H. influenzae* either in
buffer B or in buffer E (25 mM tris-HCl, pH 7.4, 5 mM 2-mercaptoethanol,
2.5 mM Na$_2$EDTA) with equal results.

After addition of an aliquot of the enzyme preparation and S-adenosyl-L-
(methyl-^{3}H)-methionine to DNA or chromatin (2 μCi was added per ml,
containing 100 μg of DNA), the reaction was done for the period of time indicated
in the description of the respective experiments.

Then, the samples were either purified from non-incorporated material on
sucrose gradients or they were deproteinized and the acid-precipitable DNA was
determined chemically (BURTON 1956) and as to the incorporation of
radioactivity as described (DOENECKE & MCCARTHY 1976).

Nuclease Digestion Experiments. Chromatin samples (in most cases
nucleosome dimers or monomers), which were either untreated or had been
modified by one of the procedures described in the other paragraphs, were
exposed to micrococcal nuclease under the following conditions. 100 units of the
enzyme were added per ml of the chromatin solution (A 260 = 2.0, unless
otherwise stated), CaCl$_2$ was added (final concentration 1 mM) and after 5
minutes of incubation at 37 °C, the reaction was stopped with 2 mM Na$_2$EDTA
(DOENECKE 1977 a).

In some of the experiments, the reactions were done in the presence of
polylysine or ethidium bromide (or both). In these cases, the concentrations of
these compounds are expressed as lysine residues per nucleotide or dye molecules
per nucleotide, respectively.

After the end of the nuclease reactions, the samples were deproteinized by
addition of NaCl (final concentration 1 M), sodium dodecylsulfate (1 %, w/v) and
an equal volume of chloroform/isoamylalcohol (24/1, v/v). After shaking and
centrifugation, this step was repeated and finally the aqueous layer was dialyzed
against double-distilled water. Finally, the solution was lyophilized and the
material was further analysed as indicated in the respective figures and in the
results section.

In experiments with radioactively methylated DNA, the samples were
precipitated with trichloroacetic acid (5 %, w/v) on filter paper (Schleicher and
Schuell, 2043 b) and counted (DOENECKE& MCCARTHY 1976).

Polyacrylamide Gel Electrophoresis. The lyophilized DNA samples after
digestion and deproteinisation were analyzed on 5 % polyacrylamide slab gels
(LOENING 1967) as described previously (DOENECKE 1977 a).

The nucleosomal proteins or total lyophilized nucleosomes were analyzed by
electrophoresis on 15 % polyacrylamide-SDS-gels according to LAEMMLI (1970).
To that purpose, total nucleosomes (i.e. DNA and proteins together) were
lyophilized or precipitated with ethanol and dissolved in 0.1 M Tris-HCl, pH 6.8,
2 % (w/v) sodium dodecylsulfate, 2 % (v/v) 2-mercaptoethanol and 20 % (v/v)
glycerol and electrophoresed on 15 % polyacrylamide gels.

The crosslinked octamers of histones were identified on 5 % polyacrylamide
SDS gels (WEBER & OSBORN, 1969) as described by STEIN & al. (1977).

The DNA gels were stained with "stains all" [= 1-ethyl-2-(3-(1-ethylnaphtho(1,2-d)thiazolin-2-ylidene)-2-methylpropenyl)-naphtho(1,2-d)thiazoliumbromide] and the proteins with amido black.

Scatchard Analysis of Dye Binding Sites. The accessibility of the nucleosomal DNA to intercalation of ethidium bromide was measured by spectrophotometric titration of nucleosome monomers. The decrease of the absorbance at 460 nm upon reaction of increasing concentrations of nucleosomal DNA with a constant amount of ethidium bromide was used for a calculation of the binding parameters (DOENECKE 1976 b) as described by WARING (1965) and the data thus calculated were plotted according to SCATCHARD (1949).

Results

Methylation of Nucleosomal DNA With Enzymes From Haemophilus influenzae.
The methylation rate of DNA in the intact chain of nucleosomes is reduced by 80% when compared with free DNA of the same DNA fragment length (Fig. 1). In this case, nucleosome dimers were chosen as a model for the chain of nucleosomes, since they contain its two main features, i.e. the nucleosomal core DNA and the connecting linker DNA segment. Longer chains of nucleosomes are methylated at a lower rate of modification than nucleosome dimers, whereas the methylation experiments with nucleosome monomers gave varied results, when compared with dimers. Nucleosome monomer cores, finally, were methylated only at a very low rate (less than 20% of the nucleosome dimer modification rate).

Nuclease Digestion of Methylated Nucleosome Dimers.
The DNA fragment pattern after digestion of methylated nucleosome dimers with micrococcal nuclease was unchanged when compared with un-methylated controls. The same is true for the interference of polylysine with the nuclease digestion (Fig. 2). At low lysine/nucleotide ratios, polylysine preferentially protects the nucleosomal core DNA (DOENECKE 1977 a) against internal cleavage, whereas the internucleosomal cleavage still occurs and nucleosome monomers are released.

Analysis of the radioactivity, which remains trichloroacetic acid precipitable after treatment with micrococcal nuclease reveals, that in the absence of polylysine, when no dimers are left, but DNA of subnucleosomal fragment size is present, the radioactive (i.e. methylated) DNA is totally degraded, whereas about 60% of the total DNA is still precipitable and can be determined chemically (Fig. 2).

With increasing protection, methylated regions are preserved, but only in such samples, where nucleosome dimers are not degraded below the nucleosomal monomer size. This indicates, that methylated DNA is present only in such cases, when the linker DNA (as in dimers) or the terminal tails (as remainders of linker DNA in nucleosome monomers) are preserved.

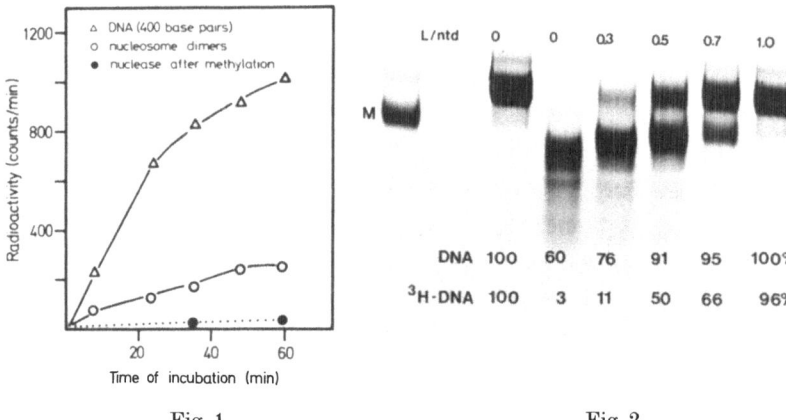

Fig. 1   Fig. 2

Fig. 1. Methylation of nucleosome dimers and DNA (protein-free, extracted from nucleosome dimers) with modification methylases from H. influenzae Rd. with S-adenosyl-L-(methyl-³H) methionine as the donor of methyl groups. After 36 and 60 min of methylation of nucleosome dimers aliquots were taken from the incubation mixture, CaCl₂ (1 mM) and micrococcal nuclease (final concentration 100 U/ml) were added and after 5 min further incubation, the reaction was stopped by chilling and addition of Na₂EDTA (2 mM). The methylated samples as well as the methylated and nuclease-treated samples were then deproteinised and precipitated with trichloroacetic acid

Fig. 2. Digestion of methylated nucleosome dimers with micrococcal nuclease. ³H-methylated nucleosome dimers were incubated in the absence and in the presence of increasing amounts of polylysine with micrococcal nuclease (100 U/ml) for 5 min at 37 °C. Polyacrylamide gel electrophoresis of deproteinized digestion products. The polylysine concentration is expressed as lysine residues/nucleotide. The values in the DNA-line represent acid-precipitable, chemically determined DNA, expressed as percent of the undigested control (first dimer slot). The ³H-DNA-line gives the values for acid-precipitable, radioactively labelled (i.e. methylated) DNA. M: nucleosomal monomer DNA as a 200 nucleotide pair marker

Thus, we conclude, that the methylation of the nucleosomal DNA with modification methylases from *H. influenzae* takes only place at the internucleosomal DNA. This is supported by the result shown in Fig. 3. Addition of polylysine at a lysine/nucleotide ratio of 0.3 interferes with the digestion by micrococcal nuclease by less than 10 %. This indicates again, that the sites of preferential interaction of micrococcal nuclease and the modification methylase with nucleosomal DNA on one hand and polylysine on the other hand are not identical. Similarly, methylation of nucleosome dimers in the presence of polylysine is again only slightly decreased in comparison with control samples (not shown).

DNase I Digestion of Methylated Nucleosomes. Digestion of methylated nucleosomes with DNase I again shows the characteristic digestion pattern with 10 base fragments and multiples of it (NOLL 1974 b) and thus proves again, that the methylation reaction *per se* does not change the general structure of the nucleosome.

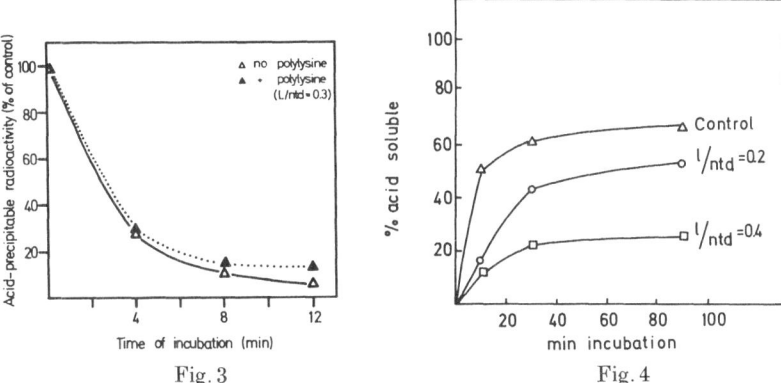

Fig. 3   Fig. 4

Fig. 3. Digestion of methylated nucleosome dimers in the absence and in the presence of polylysine (lysine/nucleotide = 0.3). Digestion conditions: 5 min, 37 °C, 50 U microcococal nuclease/ml, 1 mM $CaCl_2$, stop of reaction with 2 mM Na_2EDTA, deproteinization (see mathods) and precipitation with trichloroacetic acid

Fig. 4. Digestion kinetics of nucleosome trimers in the presence of polylysine at lysine/nucleotide ratios 0.2 and 0.4, respectively. The ordinate shows the percentage of total DNA which is rendered soluble in 0.5 M $HClO_4$, 0.5 M NaCl by micrococcal nuclease (50 U/ml) after the times of incubation given on the abscissa

A comparison of the release of radioactively methylated DNA and total DNA (determined chemically) did not reveal a difference in the kinetics of degradation of methylated and total DNA. This is consistent with the fact, that this nuclease cleaves (in contrast to micrococcal nuclease) the nucleosomal core and linker DNA portions at the same rate (NOLL 1977).

Binding of Polylysine to Nucleosomal DNA. We have shown previously (DOENECKE 1977 a) and demonstrated again in the previous paragraphs, that polylysine preferentially interacts with the nucleosomal core DNA. This was concluded from the fact, that addition of polylysine at a lysine/nucleotide ratio of 0.3, which would suffice to protect the whole internucleosomal linker DNA against cleavage by micrococcal nuclease, leads to a protection of the nucleosomal core against internal cleavage, whereas cleavage at the linker DNA site continues (see Fig. 2). The protection against degradation of the internucleosomal DNA with increasing polylysine concentration, which is demonstrated in Fig. 4 is therefore attributed to an interaction of polylysine with the nucleosomal core DNA. The binding stoichiometry of polylysine to nucleosomes (as to whole chromatin) can be determined by titration and precipitation (CLARK & FELSENFELD 1971, DOENECKE 1977 a) and upon addition of polylysine no histones are released from nucleosomes (DOENECKE 1977 a).

The artificial increase of the "protein"/DNA ratio due to polylysine binding to nucleosomes can be measured by fixation of the polylysine/ nucleosome complex with formaldehyde and its subsequent centri- fugation in cesium chloride equilibrium density gradients. Fig. 5 shows, that the density of the fixed nucleosome dimers decreases upon addition of polylysine at a lysine/nucleotide ratio of 0.2. The protein/DNA ratio, which can be calculated from the buoyant density of the peak fractions, increases proportionally.

Thus, we conclude from the gel patterns shown above and from data published previously (DOENECKE 1977 a), together with the digestion kinetics of methylated and unmodified DNA, that polylysine effectively interacts with the nucleosomal core DNA.

The nucleosomal core DNA is, however, the DNA segment, which is associated with the histone octamer. Since histones block the major groove of the chromosomal DNA (MIRZABEKOV & KOLCHINSKY 1974) and since polylysine binding does not cause the release of any histones from nucleosomes (DOENECKE 1977 a), it seemed reasonable to test, whether the minor DNA groove is the site of polylysine binding to DNA in nucleosomes. The tool for this investigation was ethidium bromide, a drug molecule, which intercalates in between the base pairs in DNA via the minor groove (TSAI & al. 1975) and needs accessible phosphate groups for intercalation (ANGERER & MOUDRIANAKIS 1972).

Ethidium Bromide Binding to Nucleosomal DNA. Competition of ethidium bromide and polylysine would give us a proof, that the site of polylysine binding to the nucleosomal core DNA is in fact the minor groove. To that purpose, two sets of experiments were done. First, the amount of binding sites accessible to ethidium bromide for intercalation were determined spectrophotometrically in the absence and in the presence of polylysine and, secondly, the effect of ethidium bromide binding on nucleosomal DNA cleavage patterns (DOENECKE 1977 b) was studied in the presence of polylysine.

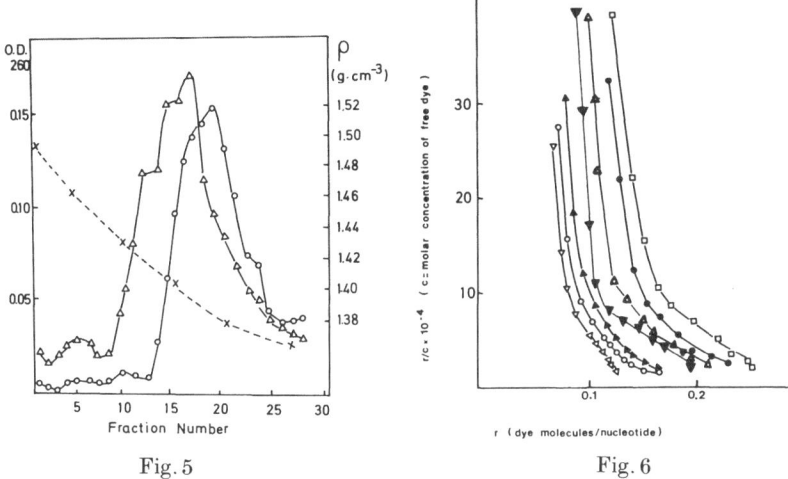

Fig. 5   Fig. 6

Fig. 5. Buoyant density of nucleosome monomers (Δ) fixed with formaldehyde and banded in cesium chloride equilibrium density gradients. Comparison with nucleosome-monomer/polylysine complexes (L/ntd = 0.2), which were equally fixed and centrifuged (O). Calculation of the increase of the "protein"/DNA ratio (BRUTLAG & al., 1969) indicates an increase of 18% in the polylysine/nucleosome complex (compared with nucleosome peak fraction). Gradient conditions: Beckman SW 60, 180.000 xg, 60 hours, 18 °C, dilution of fractions with 2 ml of H_2O for measurement of A 260

Fig. 6. Scatchard plots describing ethidium bromide binding to free DNA (□), nucleosome monomers (●), polylysine/nucleosome monomer complexes at lysine/nucleotide ratios 0.1 (Δ) and 0.2 (▼), nucleosome dimers (▲), trimers and multimers (O, ▽). r, amount of dye molecules bound per nucleotide; c, molar concentration of free dye; r and c were calculated according to Waring (1965).
— (Redrawn in part from DOENECKE 1977 b)

Ethidium bromide binds to nucleosomal DNA (Fig. 6). The accessibility of the DNA in nucleosomes is nearly as high as in free DNA (binding sites are available along nearly the entire length of the nucleosomal DNA), but when the chromatin fragment size is increased, the amount of dye binding sites per nucleotide decreases. This might be due to the fact, that in longer chains of nucleosomes the presence of non-histone proteins, higher order structures and close contacts between neighbouring nucleosome cores reduce the accessibility of the DNA to ethidium bromide and prevent a relaxation of the DNA/histone interactions. Finally, this type of spectrophotometric determinations of ethidium binding sites does not discriminate between different types of binding (LAWRENCE & DAUNE 1976).

As to the purpose of the experiments reported here, however, the data clearly show (Fig. 6), that polylysine drastically interferes with ethidium bromide intercalation. Thus, both molecules seem to compete for the minor groove of DNA.

The same conclusion can be drawn from a nuclease assay. Digestion of nucleosomal DNA with micrococcal nuclease in the presence of

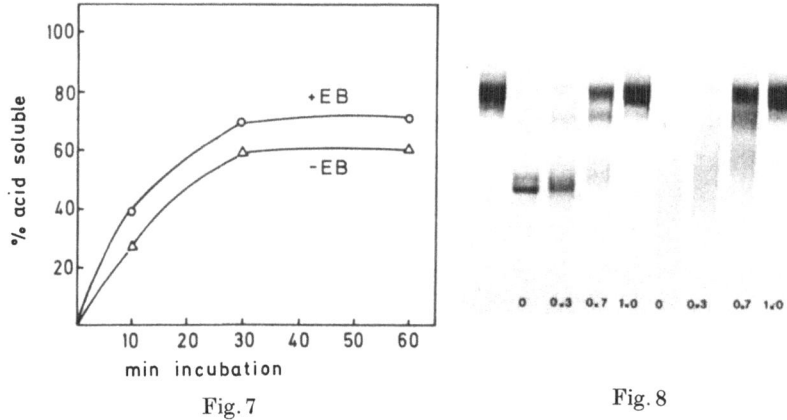

Fig. 7   Fig. 8

Fig. 7. Digestion kinetics of nucleosome monomers in the absence and in the presence of ethidium bromide (dye/nucleotide ratio 0.3). Ordinate: percentage of total DNA which is rendered soluble in 0.5 M $HClO_4$, 0.5 M NaCl. Digestion conditions: micrococcal nuclease (50 U/ml), 37 °C, 1 mM $CaCl_2$, stop at times indicated with Na_2EDTA (2 mM)

Fig. 8. Digestion of nucleosome trimers with micrococcal nuclease in the absence (slot 1-5, from left to right) and in the presence of ethidium bromide (slot 6-9, dye/nucleotide = 0.3). The numbers indicate the lysine/nucleotide ratios during the incubation with micrococcal nuclease (5 min, 37 °C, 100 U/ml, 1 mM $CaCl_2$, stop with 2 mM Na_2EDTA and deproteinization). 5% polyacrylamide slab gel, for details see Methods

ethidium bromide results in an increase of the amount of DNA, which is rendered acid-soluble (Fig. 7). We have shown previously, that this effect is accompanied by a loss of specificity of cleavage at defined nucleosomal DNA sites (DOENECKE 1977 b). In analogy to the determination of binding sites (Fig. 6), we now investigated, whether polylysine could counteract this effect of ethidium bromide on cleavage patterns of nucleosomal DNA with micrococcal nuclease, as well.

Fig. 8 reveals, that polylysine competes against ethidium bromide also in respect to its effect on cleavage patterns. Polylysine restores the specificity of cleavage (at internucleosomal sites), which had been distorted by the drug.

Cleavage of Cross-Linked Nucleosomes in the Presence of Ethidium Bromide. Ethidium intercalation changes the conformation of the DNA, and at high salt concentrations the release of histones is facilitated (BENYAJATI & WORCEL 1976, STRÄTLING & SEIDEL 1976, FENSKE & al. 1975). We have shown (Figs. 7 and 8), that this results in a change of accessible sites for micrococcal nuclease. This accessibility depends on the distribution of proteins along the DNA. Thus, the question arises, whether the distribution of proteins (i.e. the organization of the histone octamer) along the DNA is changed, when ethidium bromide alters the conformation of the DNA moiety of the nucleosome.

This question was approached in two ways. First, nucleosomes were cross-linked with formaldehyde in the absence and in the presence of ethidium bromide (at a dye/nucleotide ratio, which distorts cleavage patterns drastically) and then the nucleosomal proteins were analysed as to the formation of the characteristic multimers of histones as described by VAN LENTE & al. (1975). Fig. 9 shows, that the pattern of histone oligomers remains unchanged, when the cross-linking of histones and DNA is done in the presence of ethidium bromide (the figure also shows, that fixation in the presence of polylysine results in a cross-link pattern which is identical with the control preparation. Thus, polylysine and histones are not in a close contact and are not linked to each other).

The cross-link-patterns in the presence of ethidium bromide apparently indicate, that the change in the accessibility of the nucleosomal DNA is not accompanied by changes within the histone octamer. It should, however, be kept in mind, that the cross-linked histone oligomers, which we can resolve in these sodium dodecylsulfate polyacrylamide gels, represent only 15% of the total nucleosomal histones (VAN LENTE & al. 1975) and cross-linked DNA/histone complexes and histone multimers fail to enter the gel (DOENECKE 1978).

The second approach was based on the cross-linking of the nucleosomal core histones to each other with dimethylsuberimidate. In this case, the histone octamer is preserved throughout cleavage of the nucleosomal DNA with micrococcal nuclease and DNase I (STEIN & al. 1977) and the DNA cleavage patterns do not differ from patterns obtained with native nucleosomes. Digestion of the nucleosomal DNA in such cross-linked nucleosomes with micrococcal nuclease in the presence of ethidium bromide results, however, in a change of the pattern of the DNA digestion products. The average size of DNA fragments remains greater, even after prolonged cleavage, and some prominent DNA bands remain visible, whereas the untreated material (not cross-linked) is

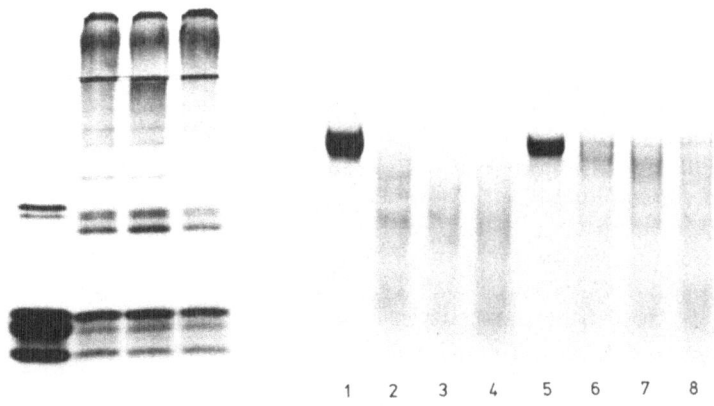

Fig. 9   Fig. 10

Fig. 9. Cross-linking of DNA and histones with formaldehyde as described by
van Lente & al. (1975). From left to right: untreated control nucleosome
monomers, formaldehyde-treated nucleosome monomers, formaldehyde-treated ethidium/nucleosome complex (dye/nucleotide at the time of fixation:
0.15), formaldehyde-treated polylysine/nucleosome complex (lysine/nuc-
leotide: 0.3). — 2 A260 units of nucleosome monomers were fixed with
formaldehyde (1%) for 24 hours at + 4 °C, dialysed against water, lyophilised
and dissolved in electrophoresis buffer (Laemmli 1970)

Fig. 10. Digestion of untreated (slots 1-4) and dimethylsuberimidate-
crosslinked (5-8) nucleosome cores with micrococcal nuclease in the absence and
presence of ethidium bromide. Slots 1 and 5: undigested controls; 2 and 6:
digestion in the absence of ethidium bromide; 3 and 7: dye/nucleotide 0.15; 4
and 8: dye/nucleotide 0.3. — Digestion conditions: micrococcal nuclease (5 min,
37 °C, 100 U/ml, 1 mM CaCl₂, stop with 2 mM Na₂EDTA, deproteinization)
Dialysis against H₂O, lyophilisation, samples dissolved in half strength
electrophoresis buffer. 5% polyacrylamide gel, stained with stains all

randomly cleaved (Fig. 10). This result could indicate, that upon
fragmentation of the DNA (in native nucleosomes) in the presence of
ethidium bromide, the affinity of the DNA-ethidium complex to a
slightly modified histone octamer is changed and allows an intense
degradation of the nucleosomal DNA. On the other hand, degradation of
the nucleosomal DNA in the presence of ethidium bromide with a cross-
linked histone octamer does not result in a change of histone/DNA
interactions to such an extent and leads to a protection of a greater DNA
fragment than with non-cross-linked nucleosomes.

Further experiments are needed and will hopefully help to resolve the discrepancies between the data obtained with formaldehyde- and dimethylsuberimidate-treated nucleosomes in the presence of ethidium bromide.

Discussion

Nucleases are the most frequently used tools for probing the structure of chromatin. Since they differ in their mode and site of action, such enzymes can be used to describe nucleosomal DNA sites, which differ in their accessibility to the various enzymes used. The description of the topography of accessible regions relies mainly on DNA fragment patterns obtained after the action of the enzymes on nucleosomes and whole chromatin preparations.

Other approaches, which describe the internal arrangement of the nucleosome core through labelling, cross-linking and digestion techniques (SIMPSON & WHITLOCK 1976, MIRZABEKOV & al. 1978) also rely on nucleases as the main tool for the analysis of the modified nucleosomes, but cleavage of DNA is in this case a tool, which helps to define the mode of action of the other analytical approach.

Thus, modification procedures, combined with established digestion techniques, help to interpret nuclease cleavage data and *vice versa* and to overcome some of the obvious disadvantages of digestion techniques, such as the fact, that the degradation of the accessible site is an inherent feature of a cleavage approach.

Methylation of Accessible Segments of the Nucleosomal DNA. Methylation of the nucleosomal DNA with modification enzymes from *Haemophilus influenzae* occurs primarily at the internucleosomal linker DNA. This can be easily proven with micrococcal nuclease, which preferentially degrades the linker DNA under mild digestion conditions. When methylated nucleosome dimers (as a model of the chain of nucleosomes) are exposed to micrococcal nuclease, the methylated DNA regions are rapidly degraded, whereas most of the total DNA remains acid-precipitable and can be resolved electrophoretically as a fragment pattern in the subnucleosomal range.

Partial protection with polylysine allows to preserve larger fragments against degradation by micrococcal nuclease. This results in a spectrum of DNA fragments in the size range between subnucleosomal fragments and nucleosome dimers. Analysis of these DNA fragments as to radioactivity (i.e. methyl groups) reveals, that fragments, which are greater than nucleosome cores remain radioactively labelled, whereas smaller fragments do not. Thus, modification is restricted to the internucleosomal DNA.

This does not mean, that the specific target nucleotide sequences of the different methylases are preferentially located at these internucleosomal linker DNA sequences, but the result rather reflects structural differences between the nucleosomal core and linker moieties. Since histones block the major groove of the nucleosomal core DNA (MIRZABEKOV & KOLCHINSKY 1974), the amino group of the adenine bases, where the modification enzymes methylate, are not accessible in the core DNA, but readily available in the linker region.

Interaction of Polylysine and the Nucleosomal Core DNA. Polylysine causes a partial protection of the chains of nucleosomes against cleavage by micrococcal nuclease. This allows a restriction of the action of this enzyme to its primary site of action, the internucleosomal linker DNA. The opposite might have been anticipated, i.e. a preferential binding of polylysine to the relatively "open" linker DNA. This latter possibility was excluded by adding enough polylysine to cover 30 % of the nucleosomal DNA (which would correspond to 60 out of 200 nucleotide pairs, i.e. the size of the linker DNA). In this case, the linker remained unprotected and subnucleosomal cleavage was inhibited.

Our interpretation of these data was supported by the fact, that the characteristic DNase I cleavage pattern of the nucleosomal core DNA was not obtained in the presence of polylysine (DOENECKE 1977 a). Furthermore, the kinetics of nucleosome methylation and of the degradation of methylated regions in the presence of polylysine support the idea, that polylysine interacts with the nucleosomal core DNA and does not interfere with reactions, which preferentially occur at the internucleosomal linker DNA.

Binding of Ethidium Bromide to Nucleosomal DNA. The nucleosomal core DNA is the segment, which is bound to histones. These proteins block the major groove of the nucleosomal DNA (MIRZABEKOV & KOLCHINSKY 1974, MIRZABEKOV & al. 1978). We therefore concluded, that the only site, at which polylysine could bind to the nucleosomal core DNA, was the minor groove. This idea was supported by model studies done by WILKINS (1956) with polyarginine and by actinomycin-D-binding studies with pentalysine and free DNA (CARROLL & BOTCHAN 1972).

We have shown, that binding of polylysine to the phosphate groups along the minor groove blocks this site against intercalation of ethidium bromide. This counteraction of polylysine against ethidium intercalation was demonstrated by determination of binding parameters and calculation of Scatchard plots (SCATCHARD 1949) and by a reversal of the effect of ethidium bromide on nucleosomal DNA cleavage patterns (DOENECKE 1977 b).

Since changes in cleavage patterns suggest, that the interaction of histones and DNA has changed, we concluded, that the conformational change of the nucleosomal DNA due to ethidium intercalation led to slight changes in the mode of interaction of the nucleosomal DNA and

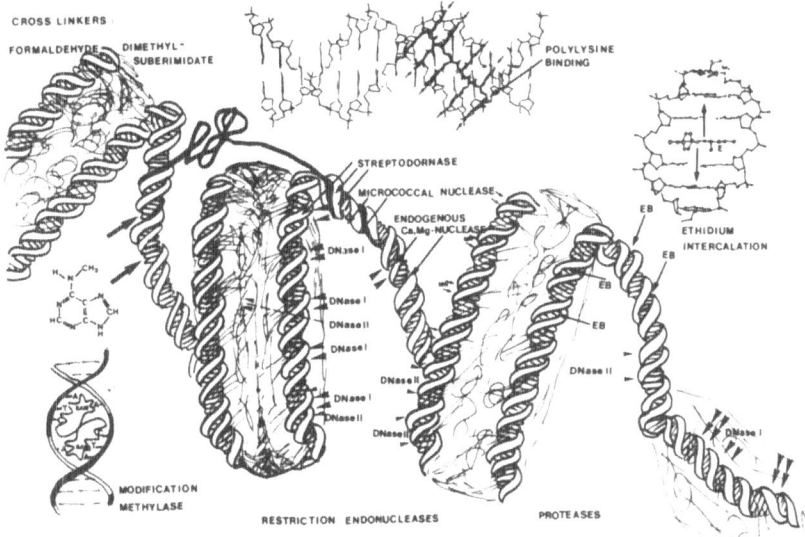

Fig. 11. Schematic illustration of the tools described in this paper. Distribution of the sites of preferential attack by nucleases at the nucleosomal core and linker DNA, together with schemes of the mode of action of modification methylases (redrawn from BOYER 1971), polylysine (in analogy to WILKINS 1956) and ethidium bromide (TSAI & al. 1975). — The drawing does not include data on the arrangement and mode of binding of histones nor does it consider models about the superhelical arrangement of the nucleosomal core DNA, which influences the frequency of DNase I cutting sites. The hypothetical non-nucleosomal arrangement of histones and extended DNA (right, lower corner) is drawn to indicate a high accessibility of actived chromatin regions to DNase I. The terms *proteases* and *restriction endonucleases* at the bottom of the drawing indicate further procedures frequently used in chromatin studies but omitted in this paper

the histone core. Such a weakening of histone/DNA interactions can be demonstrated by a release of histones at elevated ionic strength (BENYAJATI & WORCEL 1976, STRÄTLING & SEIDEL 1976, FENSKE & al. 1975). Our experiments with cross-links between histones and histones or between histones and DNA could not finally settle the question, whether the conformational change of the ethidium-bound nucleosomal DNA was accompanied by slight changes of histone-histone contacts in the nucleosomal histone octamer. Our data suggest, that cleavage with micrococcal nuclease in the presence of ethidium bromide leads to changes in the histone/DNA contacts, which promote cleavage towards smaller DNA fragments. Crosslinking of the histones with dimethylsuberimidate prevents this effect in part.

In summary, we have demonstrated (Fig. 11), that the established nuclease digestion studies, which are of central importance for the analysis of the nucleosome structure, can be supplemented with methods, which act through modification, protection and distortion of the nucleosomal DNA. Together with cross-linking procedures, these methods should help to identify accessible regions and allow model studies about possible structural changes within the nucleosome. Hopefully, the methods will also allow a further description of differences between chromatin portions of different function *in vivo*.

This work was supported by a grant from the *Deutsche Forschungsgemeinschaft*. The technical assistance of Ms. Susanne Boos is gratefully acknowledged.

References

Altenburger, W., Hörz, W., Zachau, H. G., 1976: Nuclease cleavage of chromatin at 100 nucleotide intervals. — Nature (London) **264**, 517—522.

Angerer, L. M., Moudrianakis, E. N., 1972: Interaction of ethidium bromide with whole and selectively deproteinized deoxynucleoproteins from calf thymus. — J. Mol. Biol. **63**, 505—521.

Benyajati, C., Worcel, A., 1976: Isolation, characterization and structure of the folded interphase genome of *Drosophila melanogaster*. — Cell **9**, 393—407.

Boyer, H. W., 1971: DNA restriction and modification mechanisms in bacteria. — Annu. Rev. Microbiol. **25**, 153—176.

Brutlag, D., Schlehuber, C., Bonner, J., 1969: Properties of formaldehyde-treated nucleohistone. — Biochemistry 8, 3214—3218.

Burton, K., 1956: A study of the conditions and mechnism of the diphenylamine reaction for the colorimetric estimation of deoxyribonucleic acid. — Biochem. J. **62**, 315—321.

Camerini-Otero, R. D., Sollner-Webb, B., Simon, R. H., Williamson, P., Zasloff, M., Felsenfeld, G., 1978: Nucleosome structure, DNA folding and gene activity. — Cold Spring Harbor Symp. Quant. Biol. **42**, 57—75.

Carroll, D., Botchan, M. R., 1972: Competition between pentalysine and actinomycin D for binding to DNA. — Biochem. Biophys. Res. Commun. **46**, 1681—1687.

Chauveau, J., Moule, Y., Rouiller, C., 1956: Isolation of pure and unaltered liver nuclei. Morphology and biochemical composition. — Experim. Cell Res. **11**, 317—324.

Clark, R. J., Felsenfeld, G., 1971: Structure of chromatin. — Nature New Biology **229**, 101—106.

CLIMENT, F., DOENECKE, D., BEATO, M., 1977: Properties of the partially purified activated glucocorticoid receptor of rat liver. Binding to chromatin subunits. — Biochemistry 16, 4694—4703.

DAVIES, G. E., STARK, G. R., 1970: Use of dimethylsuberimidate, a cross-linking reagent, in studying the subunit structure of oligomeric proteins. — Proc. Natl. Acad. Sci. (U.S.) 66, 651—656.

DOENECKE, D., 1976 a: Cesium chloride gradients of chromatin after treatment with micrococcal nuclease. — Cell 8, 59—64.

— 1976 b: Binding of ethidium bromide to fractionated chromatin. — Experim. Cell Res. 100, 223—227.

— 1977 a: Binding of polylysine to chromatin subunits and cleavage by micrococcal nuclease: a comparison of accessible sites. — Eur. J. Biochem. 76, 355—363.

— 1977 b: Ethidium bromide binding to nucleosomal DNA: effects on DNA cleavage patterns. — Experim. Cell Res. 109, 309—315.

— 1978: Digestion of chromosomal proteins in formaldehyde treated chromatin. — Hoppe-Seylers Z. Physiol. Chem. 359, 1343—1352.

— MCCARTHY, B. J., 1975: Protein content of chromatin fractions separated by sucrose gradient fractionation. — Biochemistry 14, 1366—1372.

— — 1976: Movement of histones in chromatin induced by shearing. — Eur. J. Biochem. 64, 405—411.

FENSKE, H., EICHHORN, I., BÖTTGER, M., LINDIGKEIT, R., 1975: Evidence of altered histone interactions, as investigated by removal of histones, in chromatin isolated from rat liver nuclei by a conventional method. — Nucl. Acids Res. 2, 1975—1985.

FINCH, J. T., NOLL, M., KORNBERG, R. D., 1975: Electron microscopy of defined lengths of chromatin. — Proc. Natl. Acad. Sci. (U.S.) 72, 3320—3322.

— LUTTER, L. C., RHODES, D., BROWN, R. S., RUSHTON, B., LEVITT, M., KLUG, A., 1977: Structure of nucleosome core particles of chromatin. — Nature (London) 269, 29—36.

FRANKE, W. W., SCHEER, U., 1978: Morphology of transcriptional units at different states of activity. — Phil. Trans. R. Soc. Lond. B. 283, 333—342.

GAREL, A., AXEL, R., 1976: Selective digestion of transcriptionally active ovalbumin genes from oviduct nuclei. — Proc. Nat. Acad. Sci. (U.S.) 73, 3966—3970.

GOTTESFELD, J. M., 1978: Organization of transcribed regions of chromatin. — Phil. Trans. R. Soc. Lond. B. 283, 343—357.

HEWISH, D. R., BURGOYNE, L. A., 1973: Chromatin Substructure. The digestion of chromatin DNA at regularly spaced sites by a nuclear deoxyribonuclease. — Biochem. Biophys. Res. Commun. 52, 504—510.

KATO, Y., and IWAI, K., 1977: DNA-binding segments of four histone sequences identified in trypsin-treated H 1-depleted chromatin. — J. Biochem. 81, 621—630.

KORNBERG, R. D., 1977: Structure of chromatin. — Ann. Rev. Biochem. 46, 931—954.

— THOMAS, J. O., 1974: Chromatin structure—oligomers of the histones. — Science 184, 865—868.

LAEMMLI, U. K., 1970: Cleavage of structural proteins during the assembly of the head of bacteriophage T 4. — Nature (London) 227, 680—685.

LAWRENCE, J. J., DAUNE, M., 1976: Ethidium bromide as a probe of conformational heterogeneity of DNA in chromatin. The role of histone H 1. — Biochemistry 15, 3301—3307.

Loening, U. E., 1967: The fractionation of high-molecular weight ribonucleic acid by polyacrylamide gel electrophoresis. — Biochem. J. **102**, 251—257.

Mirzabekov, A. D., Kolchinsky, A. M., 1974: Localization of chromatin proteins within DNA grooves by methylation of chromatin with dimethyl sulphate. — Mol. Biol. Rep. **1**, 379—384.

— Shick, V. V., Belyausky, A. V., Karpov, V. L., Bavykin, S. G., 1978: The structure of nucleosomes: the arrangement of histones in the DNA grooves and along the DNA chain. — Cold Spring Harbor Symp. Quant. Biol. **42**, 149—155.

Noll, H., 1967: Characterization of macromolecules by constant velocity sedimentation. — Nature (London) **215**, 360—363.

Noll, M., 1974 a: Subunit structure of chromatin. — Nature (London) **251**, 249—252.

— 1974 b: Internal structure of the chromatin subunit. — Nucl. Acids Res. **1**, 1573—1578.

— 1977: DNA folding in the nucleosome. — J. Mol. Biol. **116**, 49—71.

— Thomas, J. O., Kornberg, R. D., 1975: Preparation of native chromatin and damage caused by shearing. — Science (Wash.) **187**, 1203—1206.

Olins, A. L., Olins, D. E., 1974: Spheroidal chromatin units (ν-bodies). — Science (Wash.) **183**, 330—332.

Oudet, P., Gross-Bellard, M., Chambon, P., 1975: Electron microscopic and biochemical evidence that chromatin structure is a repeating unit. — Cell **4**, 281—300.

Roy, P. H., Smith, H. O., 1973: DNA methylases of *Haemophilus influenzae* Rd. — J. Mol. Biol. **81**, 427—444.

Sahasrabuddhe, C. G., van Holde, K. E., 1974: The effect of trypsin on nuclease-resistant chromatin fragments. — J. Biol. Chem. **249**, 152—156.

Scatchard, G., 1949: The attraction of proteins for small molecules and ions. — Ann. N.Y. Acad. Sci. **51**, 660—672.

Simpson, R. T., Whitlock, J. P. Jr., 1976: Mapping DNAase I-susceptible sites in nucleosomes labeled at the 5′ ends. — Cell **9**, 347—353.

Sollner-Webb, B., Melchior, W. Jr., Felsenfeld, G., 1978: DNase I, DNase II and staphylococcal nuclease cut at different, yet symmetrically located, sites in the nucleosome core. — Cell **14**, 611—627.

Stein, A., Bina-Stein, M., Simpson, R. T., 1977: A cross-linked histone octamer as a model of the nucleosome core. — Proc. Nat. Acad. Sci. (U.S.) **74**, 2780—2784.

Strätling, W. H., Seidel, I., 1976: Relaxation of chromatin structure by ethidium bromide binding: determined by viscometry and histone dissociation studies. — Biochemistry **15**, 4803—4809.

Tsai, C. C., Jain, S. C., Sobell, H. M., 1975: X-ray crystallographic visualization of drug-nucleic acid intercalative binding. — Proc. Nat. Acad. Sci. (U.S.) **72**, 628—632.

van Lente, F., Jackson, J. F., Weintraub, H., 1975: Identification of specific crosslinked histones after treatment of chromatin with formaldehyde. — Cell **5**, 45—50.

Waring, M., 1965: Complex formation between ethidium bromide and nucleic acids. — J. Mol. Biol. **13**, 269—282.

Weber, K., Osborn, M., 1969: The reliability of molecular weight determination by dodecyl sulfate-polyacrylamide gel electrophoresis. — J. Biol. Chem. **244**, 4406—4412.

WEINTRAUB, H., GROUDINE, M., 1976: Transcriptionally active and inactive conformations of chromosomal subunits. — Science (Wash.) **193**, 848—856.

— PALTER, K., VAN LENTE, F., 1975: Histones H 2a, H 2b, H 3 and H 4 form a tetrameric complex in solutions of high salt. — Cell **6**, 85—110.

WILKINS, M. H. F., 1956: Physical Studies of the molecular structure of deoxyribose nucleic acid and nucleoprotein. — Cold Spring Harbor Symp. Quant. Biol. **21**, 75—90.

Address of the author: Prof. Dr. DETLEF DOENECKE, Institut für Physiologische Chemie der Philipps-Universität, Lahnberge, D-3550 Marburg/Lahn, Federal Republic of Germany.

Pl. Syst. Evol., Suppl. 2, 221—233 (1979)

Department of Biology, The University
of Kaiserslautern, Federal Republic of Germany

Types of Chromatin Organization in Plant Nuclei

By

Walter Nagl and **Hans-Peter Fusenig**, Kaiserslautern

Key Words: Angiosperm systematics, nuclear ultrastructure, chromatin condensation, DNA contents.

Abstract: The nuclear ultrastructure of 15 angiospermal plants was studied with respect to the structural type, the proportion of condensed chromatin, the diameter of the chromatin fibers, and the DNA content. It was found that (1) at least three species-specific ultrastructural types can be distinguished among plants: the diffuse, the chromomeric and the chromonematic type: the diffuse nuclei exhibit always chromocenters, but evenly distributes, often hardly visible euchromatin, the two other types occur in sub-types, characterized by the presence or absence of chromocenters, and exhibit euchromatin of the same electron density as heterochromatin;—(2) the species-specific nuclear type is mainly determined by the nuclear DNA content (2 C value), high values favoring the chromonematic type, low values the other three types and the appearance of chromocenters;—(3) the diameters of the chromatin fibers are similar, although significantly different between certain species, in all structural types, falling into distinct classes of about 50, 100, 150, 200, 300 and 400 Å. The proportion of the various fibers varies between species, and one or the other class may be absent in a given species;—(4) The proportion of chromatin that is condensed into an electron-dense state is positively correlated with the DNA content; while in chromocentric nuclei the condensed chromatin corresponds to heterochromatin at the light microscope level, it corresponds to both eu- and heterochromatin in the other types of nuclei.

The ultrastructure of plant cell nuclei differs significantly among species, forming a few structural types. The two extreme situations are (1) nuclei in which nearly all the DNA is concentrated in heterochromatin, in so-called chromocenters, leaving a very transparent euchromatic or even chromatin-free background, and (2) nuclei in which nearly all the DNA is located in electron-dense strands of euchromatin, the so-called chromonemata or chromatin reticulum.

The differences in nuclear structure can be attributed to differences in the extent to which the chromosomes decondense in the course of entering interphase from mitosis. Since 1934 several attempts have been made to classify the nuclear structures at the light microscope level (Eichhorn 1934, Doutreligne 1939, Tschermak-Woess 1963, Barlow 1977), but no systematic study was made at the ultrastructural level.

Little is known about the determinants and the functional significance of the different nuclear types which are found in plants. Therefore we studied the quantitative relationships between nuclear DNA content and chromatin ultrastructure, the diameter of chromatin fibers, and the proportion of condensed and decondensed chromatin in various species and under various cellular and physiological conditions. This first report deals with a quantitative analysis of species-specific nuclear structure in some higher plants, as well as their possible determinants. For convenience, we arrange the various types of nuclear ultrastructure according to their structural complexity.

Material and Methods

The plants studied are listed in Table 2. They were either collected in the field or taken from the Botanical Garden of the Kaiserslautern University. Voucher specimens or preserved in the herbarium of this University. In most cases the meristem of the embryo or of root tips was used to find comparable developmental conditions.

The material was fixed with 6.25 % glutaraldehyde in phosphate buffer, pH 7.3, for 1 h, postfixed with 1% osmium tetroxide, prestained with 1% uranyl acetate in 70% ethanol, and embedded according to Spurr (1969). The sections were stained with uranyl acetate and lead citrate and examined with a Zeiss EM 10 electron microscope.

The relative optical density of the nucleoplasm and the chromatin was measured with a Zeiss integrating photometer, type PMQ-2, on electron microscopic negative films. If the electron micrographs are developed under standard conditions and, in addition, the values are corrected for the grey level of the background, comparable extinction values are obtained as an indicator of the compactness of the chromatin and nuclei in general.

The percentage of condensed chromatin was obtained statistically by determining the area of the nuclei and of the condensed chromatin with the aid of a compensation polarplanimeter (R. Ott, type 30115). This instrument allows to define the area by driving along the outline of respective areas. The results were verified by weighing the clipped nuclei and heterochromatin, respectively.

The significance of the differences between mean fiber diameters were statistically tested by the F- and t-tests.

Diffuse Nuclei With Chromocenters

The most simple structure is represented by nuclei displaying only patches of condensed chromatin or chromocenters i.e. heterochromatin, while the euchromatin is decondensed to such an extent that it is

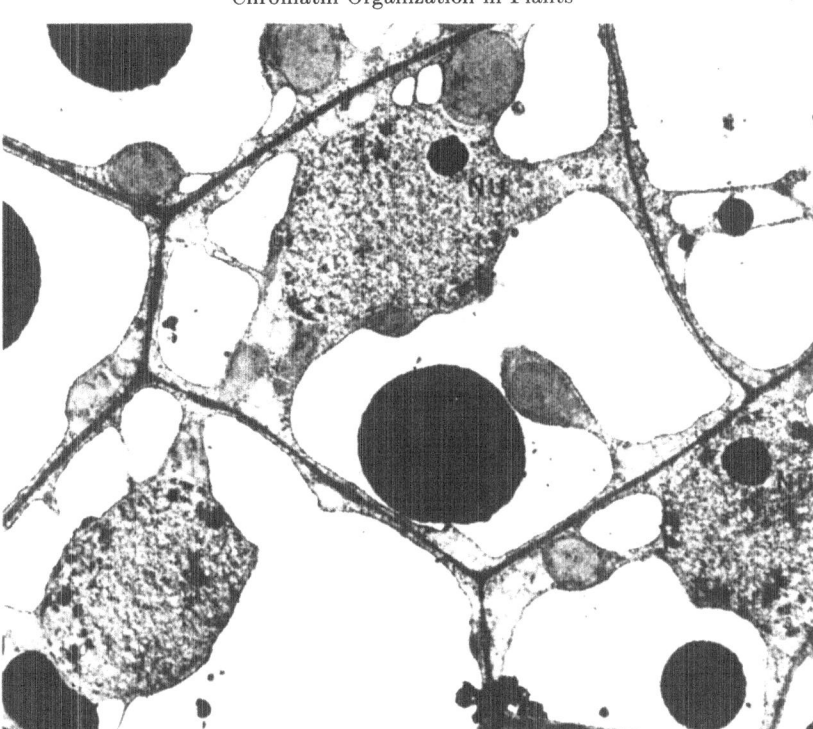

Fig. 1. *Liriodendron tulipifera*, leaf bud meristem showing nuclei of the diffuse type with some small chromocenters. NU = nucleolus (4700 ×)

indistinguishable from the nuclear matrix or appears in the form of loosely arranged tiny fibrils. This type of nuclei is termed "chromosome nuclei", "prochromosome nuclei", "chromocenter type nuclei" or "areticulate nuclei" by light microscopists. At the electron microscope level we suggest the term "diffuse nuclei". Such nuclei always display chromocenters, and most of the chromatin is in the condensed state. The size of the chromocenters may, however, vary considerably. Fig. 1 shows nuclei of the primitive angiosperm, *Liriodendron tulipifera*: they show a diffuse euchromatic structure and several small patches of heterochromatin. Fig. 2 shows a nucleus from *Rhinanthus minor* with one large chromocenter hit in this section (a complete nucleus exhibits 2 to 3 chromocenters). The rest of the nuclear cavity seems to be nearly empty. Other species which display diffuse nuclei with chromocenters are, for instance, *Beta vulgaris, Cucumis sativus, Daucus carota, Phaseolus vulgaris*, and *Sinapis alba*. In most cases all species of a genus show the same nuclear type, and often also all genera belonging to the same family.

Chromomeric Nuclei

The second type shows areas of condensed euchromatin. Such euchromatic knobs are well known from the light microscope as

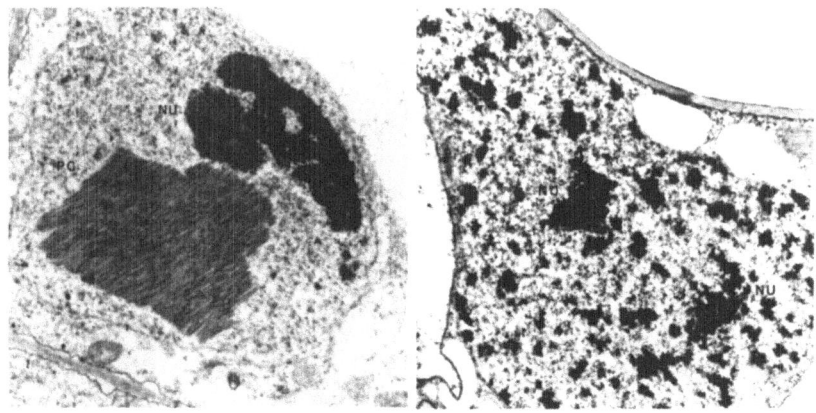

Fig. 2 (left). *Rhinanthus minor*, nucleus of a cell of the embryo. Structure is of the diffuse type with large chromocenters (only one chromocenter is visible in this section). NU = nucleolus, PC = protein crystal (7800 ×)

Fig. 3 (right). *Tectaria decurrens*, prothallium, showing a chromomeric nucleus. NU = nucleolus (7500 ×)

chromomeres, which are connected by interchromomeric linkers (the latter are only visible at pachytene and in the special case of polyteny). According to the terminology used particularly by French cytologists, nuclei of this type are semi-reticulate and reticulate, i.e. one can see more or less distinct arrays of chromomeres which may fuse to fibrils in certain regions. Chromomeric nuclei display, in ultrathin sections, small condensed patches of chromatin of the same electron density as the heterochromatin, if present, and then both types of chromatin can only be distinguished by the different size of the patches (see NAGL 1977 for other examples). In the chromomere type of nuclei the euchromatin is decondensed after mitosis to only such an extent that a certain number of chromomeres is visible, while the rest of the chromatin does not exhibit higher orders of structural arrangements. Fig. 3 displays a nucleus of *Tectaria decurrens* with a chromomeric structure which is characteristic for many ferns. Angiosperm species which display such a structure are *Pisum sativum*, *Vicia faba*, *Oxalis acetosella*, *Nicotiana tabacum* and related species and genera. Many chromomeric nuclei possess, in addition, some small chromocenters.

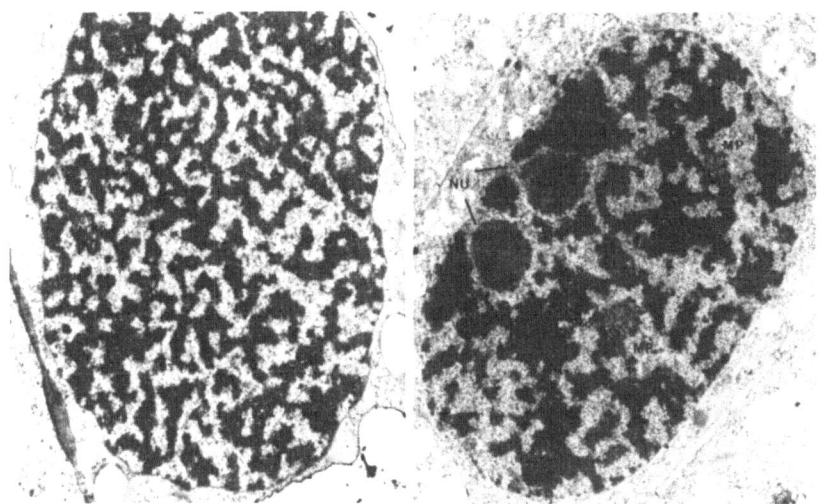

Fig. 4 (left). *Tradescantia virginiana*, root tip meristem. Typical chromonematic nucleus. The chromocenters are nearly indistinguishable from the euchromatin (5600 ×)

Fig. 5 (right). *Allium carinatum*, root tip meristem. Nucleus of highly complex organization: Structure is of the chromonematic type with large chromocenters. NU = nucleolus, MP = "micropuff" (6100 ×)

Chromonematic Nuclei

Nuclei of this type display, at the light microscope level, strands of euchromatin tortously packed in the nuclear cavity. Heterochromatin may be also present in the form of chromocenters, thus building up in conjunction with euchromatic strands, the most complex structure which is found among nuclei of meristematic plant cells. The chromonematic structure of euchromatin is the result of a low degree of its decondensation during interphase, so that strands of 0.2 to 0.5 μm in diameter can be distinguished depending on the cell cycle stage. According to the French nomenclature such nuclei are termed "reticulate" or "eureticulate", because the strands apparently form a network after fixation. At the electron microscope level the chromonematic nuclei display considerable structural changes during the cell cycle, as will be shown in a forthcoming report (for results obtained in *Allium carinatum* see Nagl 1977). Very loose strands are found in nuclei of *Juglans regia*, slightly more compact strands are seen in *Arum maculatum*. All species which have the characteristic chromonematic ultrastructure belong to the monocotyledons: *Hyacinthus orientalis*, *Muscari racemosum*, *Rhoeo discolor*, and *Tradescantia virginiana* (Fig. 4). Many species show, in addition, chromocenters of the same electron density as the euchromatic strands. In the genus *Allium*, for example, species are found with very small chromocenters and also such with very prominent ones (as in *Allium carinatum*, Fig. 5). It must be emphasized that in these complex nuclei both the heterochromatin and the euchromatin appear in a condensed state at interphase, with respect to the assembling of individual chromatin fibers into structures of higher orders. For this reason, all the electron-dense chromatin has been erroneously designated as "heterochromatin" in many publications, while the space between the strands was termed "euchromatin" (see also the discussion by Nagl 1979).

Chromatin Fibril Diameter

In order to analyze the reasons and determinants for the different nuclear structures, the chromatin is to be studied at several levels. At first we have compared the diameter of the chromatin fibrils in sectioned nuclei with various degrees of chromatin condensation. Fig. 6 shows frequency diagrams of fibril diameters as found in the nuclei of four species, two of the diffuse type with chromocenters (*Rhinanthus minor* and *Tropaeolum majus*), one of the chromomeric type (*Oxalis acetosxella*), and one of the chromonematic type (*Tradescantia virginiana*). It can be clearly seen that the chromatin fiber diameters are similar in all species studied irrespectively of the structural type, and that fiber diameters show a similar range in both euchromatin and heterochromatin, although with different frequencies. Thus, the various structural organization of plant cell nuclei is not caused, or accompanied, by considerably different degrees of DNA packaging into the chromatin fibers. Nevertheless, the mean diameters of the chromatin fibers show some significant differences between species (Table 1).

Chromatin Density and Proportion of Condensed Chromatin

The density of the nuclei was estimated by means of relative extinction values as measured photometrically on negatives of electron micrographs, which have been processed under standardized conditions.

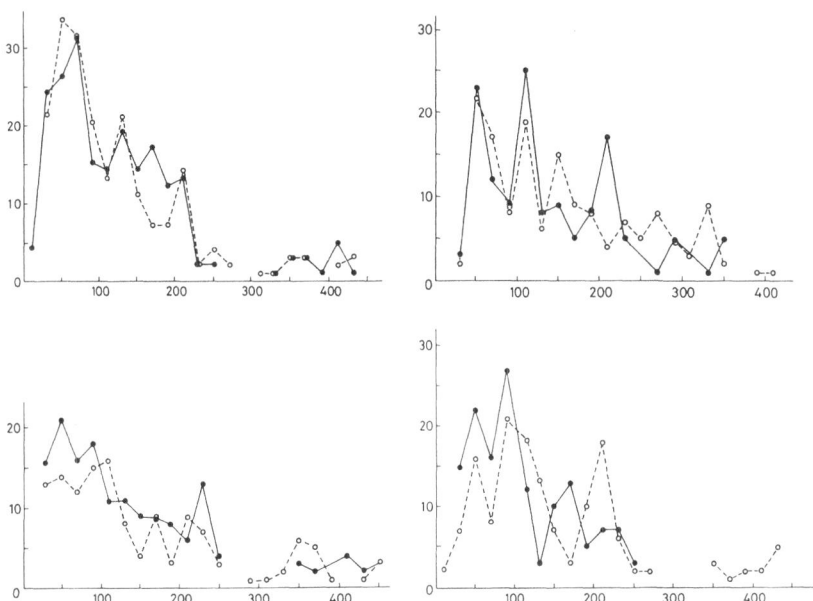

Fig. 6. Frequency diagrams illustrating the occurrence of chromatin fibers of various diameters in diffuse/chromocentric nuclei of *Tropaeolum majus* (*a*), and *Rhinanthus minor* (*b*), in chromomeric nuclei of *Oxalis acetosella* (*c*), and in chromonematic nuclei of *Tradescantia virginiana* (*d*). Note that fibers of nearly every diameter are found in the euchromatin (— — —) and heterochromatin (———) of all species

Table 1. Mean diameters of chromatin fibers as found in meristematic nuclei of plant cell nuclei belonging to different structural types, and significance of the differences of means of the same class between species[a]

Species	Mean chromatin Fiber diameters and Statistical significance of differences				
	A	B	C	D	E
·*Tropaeolum majus*	19.6 ⌐	46.8 ⌐	88.2 ⌐	175.0 ⌐	384.5 ⌐
Rhinanthus minor	27.1 ⌐	58.0 ⌐	105.2 ⌐	210.8 ⌐	324.9 ⌐
Oxalis acetosella	27.8 ⌐	49.7	102.0	194.1	388.5 ⌐
Tradescantia virginiana	21.6	48.8 ⌐	101.1	198.1	407.2 ⌐

[a] The brackets indicate significance at least at the 5% level (t-test).

15*

It was found that the optical density of the nucleoplasm and of the chromatin (chromocenters, chromomeres, chromonemata) are positively correlated (Fig. 7). Moreover, both the density of the nucleoplasm and of the chromatin increases with increasing nuclear DNA content. This indicates that the nuclear volume does not increase in the same fashion as the nuclear DNA content, but to a lesser extent.

The percentage of chromatin that is found in a condensed state is also positively correlated with the 2 C value of the species studied. In

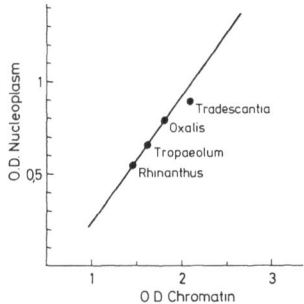

Fig. 7. Diagram illustrating the relationship between the relative optical density (O. D.) of the nucleoplasm and of the chromatin in four plant species. Note that the O. D. of both is also positively correlated with the nuclear DNA content (see Table 2)

Tropaeolum and *Rhinanthus* respectively 9.0 and 10.8 % of the nuclear area are occupied by condensed chromatin (i.e., in these species, heterochromatin). In *Oxalis* and *Tradescantia*, where also the euchromatin appears condensed into chromomeres and chromonemata, respectively, about 39.5 and 48.2 % of the nuclear area are occupied, on average, by electron-dense chromatin.

Nuclear DNA Content

The next question was as whether the basic nuclear DNA content (2 C value, i.e. the DNA content of diploid nuclei in the G_1 stage of the mitotic cell cycle) may influence the nuclear structure, as suggested earlier (Lafontaine 1974, Barlow 1977, Nagl 1977). Therefore, the 2 C values of the species studied were either taken from the literature, or measured by scanning cytophotometry, and compared with the species-specific nuclear structure (Table 2). If species are arranged according to their DNA content and nuclear complexity, a clear relationship can be seen (Fig. 8): The higher the 2 C value, the higher is the structural

Table 2. Relationship between nuclear DNA content (2C) and nuclear ultrastructure

Species	Class[a]	DNA (pg)	Reference[b]	Structure of Euchromatin[c]	Chromo-centers[d]
Allium carinatum L. (3 ×)	M	32.67	(×)	chromonematic	+
Arum maculatum L.	M	21.8	(1)	chromonematic	—
Brassia maculata R. Br.	M	7.05	(2)	diffuse	+
Clivia miniata REGEL	M	35.3	(1)	chromonematic	—
Cymbidium pumilum ROLFE	M	8.7	(2)	diffuse/"empty"	+++
Helianthus annuus L. cv. 'Russ. M'	D	8.9	(3)	diffuse	++
Hordeum vulgare L. (cv. 'Sultan')	M	11.1	(1)	chromonematic	—
Hyacinthus orientalis L.	M	49.7	(1)	chromonematic	—
Liriodendron tulipifera L.	D	1.58	(4)	diffuse	+
Oxalis acteosella L.	D	6.4	(×)	chromomeric	—
Phaseolus coccineus L.	D	3.5	(1)	diffuse	++
Rhinanthus minor L.	D	7.88	(×)	diffuse/"empty"	+++
Rhoeo discolor (L'HÉRIT.) HANCE ex WALP.	M	14.4	(1)	chromonematic	—
Tectaria decurrens (K. B. PRESL) COPEL.	F	?	—	chromomeric	++
Tradescantia virginiana L.	M	56.5	(×)	chromonematic	++
Tropaeolum majus L.	D	2.7	(5)	diffuse/"empty"	+++

[a] D = dicot. F = fern. M = monocot. [b] x = value determined in this investigation. 1 = Bennett and Smith, 1976; 2 = NAGL & CAPESIUS 1977; 3 = NAGL & CAPESIUS 1976; 4 = NAGL & al. 1977; 5 = NAGL 1976a). [c] "empty" means that no euchromatic structure is identifiable; [d] + = dominant, + + + = present, — = absent or very small

complexity. In general, monocots more frequently show high 2 C values and high nuclear complexity than dicots.

From these data we conclude that the basic nuclear DNA content is one of the determinants of the species-specific nuclear structure and its complexity.

Discussion

In this paper we made the attempt to characterize and classify the various types of nuclear ultrastructure as found in meristematic plant cells. The main results are the following: (1) The nuclear ultrastructure is

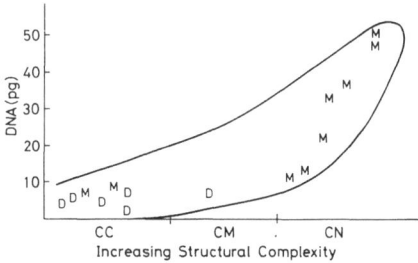

Fig. 8. Diagram to illustrate the relationship between the nuclear DNA content (2 C value) and the structural complexity of the nuclei (species are taken from Table 2).   D = dicot,   M = monocot,   CC = diffuse/chromocenter   type, CM = chromomeric type, CN = chromonematic type

species-specific; (2) the nuclear structure exhibits various degrees of complexity with respect to higher orders of chromatin condensation, while the chromatin filaments show nearly the same classes of diameters in all species; (3) the complexity of the species-specific nuclear structure is positively correlated with the basic nuclear DNA content (2 C value). While the species-specificity was already suggested by earlier investigators on the basis of nuclear types as seen in the light microscope (reviewed by Tschermak-Woess 1963), only little was known so far about the arrangement of the chromatin fiber at the electron microscope level, and only the chromocentric structure was distinguished from the reticular structure (Lafontaine 1974).

The main question which arises in comparing the differences in nuclear organization is that of its determinants. Dayal (1975) reported that imbred lines of radish show more chromocenters than the original heterozygote population or F_1 interlinear hybrids. This indicates that this nuclear character is genotypically controlled. The positive correlation between nuclear complexity and nuclear DNA content gives, however, evidence for some nucleotypic effect of the DNA mass on chromatin structure. The nucleotype (BENNETT 1973) is well known to influence nuclear and chromosome size, cell cycle duration and other parameters such as the interspersion pattern of unique and repetitive DNA sequences (reviewed by NAGL 1978). As the nucleotypic effects are independent of the nucleotide sequence, others than the genetically informative DNA sequences should be the determinants of nuclear organization. Actually some evidence has been found that the repetitive DNA plays the main role in this direction (NAGL 1976 b, 1977, 1979, BARLOW 1977).

Taken our present and the earlier findings together, the species-specific nuclear structure can be understood as the result of the following factors:

(1) Low 2 C values lead to a diffuse euchromatin organization and to the appearance of both heterochromatin and satellite DNA (see also INGLE & al. 1975, BARLOW 1977), while 2 C values higher than 9 pg lead to chromomeric or chromonematic organization. The highest DNA values are found in chromonematic nuclei which also show heterochromatin (chromocenters). As it is primarily the non-coding repetitive DNA fraction that varies among species, it can be concluded that this fraction is the main reason for the variation of the structural complexity of the nuclei.

(2) As shown by the relative optical density (extinction) of chromatin in ultrathin sections of various types of cell nuclei, the overall packaging of chromatin is different in diffuse, chromomeric and chromonematic nuclei, increasing in this order. This is consistent with the finding of BARLOW (1977) that "areticulate" nuclei seem to have a lower amount of DNA per unit volume than "reticulate" nuclei.—In more complex nuclei, the density of euchromatin and heterochromatin is identical, although it cannot be excluded that some puffed euchromatin was overlooked, because of a density identical with that of the nuclear sap.

While it is evident that the DNA content (2 C value) plays an important role in the determination of the nuclear structural organization, the *mechanism* remains unclear. In order to understand the relationships between the various levels of chromatin condensation it is also necessary to study the chromosomal proteins, from the level of the nucleosome to the level of chromonema and chromocenters. At present, we can state that the nuclear structure is rather independent of supercoiling of the nucleosome fibril *per se*, because similar fiber diameters were found in the nuclei of all species studied irrespectively of their DNA content and structure, although they show small, but significant differences between certain species. Moreover, the fibrils in euchromatin and heterochromatin are essentially of the same diameter, although thicker fibers are more frequently found in heterochromatin. These results are different from the observation of OLSZEWSKA (1973) in extended and condensed interphase and in metaphase chromatin of *Cucurbita pepo*, a plant with a low nuclear DNA content, and a DNA-rich species, *Haemanthus katharinae*, where fiber diameters are reported of 2.3 nm in extended chromatin, but 3.0 nm in condensed chromatin (the independence from the nuclear DNA amount was, however, also noted by this author). We conclude that OLSZWESKA (1973) has emphasized her study on the most common fibers that occurred in the nuclei and therefore found the differencies. The analysis of the dominant fibers may have led to the suggestion of different fiber diameters in euchromatin and heterochromatin also in other organisms (e.g. *Dipodomys*; COMINGS & OKADA 1976). The understanding of chromatin organization beyond the level of fibrils evidently requires the three-dimensional analysis and reconstruction of possible solenoids, tubes, domaines, chromosome scaffolds and other structures.

Only a few attemps have been made so far to correlate the structural organization of the nucleus to the phylogenetic position of a species. TANAKA (1971) described five different types of nuclei in the *Orchidaceae* and noted that samples of the taxonomic groups with large variation in external morphology and rapid speciation, e.g. *Orchis, Dendrobium, Oncidium, Vanda* and their allied genera, show a complex euchromatic structure with several chromocen-ters, while members of most of the taxonomic groups which show small variation and seem to be stable and slowly speciating display the diffuse type or the prochromosome (i.e. chromocentric) type. In general, progressive taxa show complex nuclear organization, more stable taxa a rather simple organization. NAGL & CAPESIUS (1977) found similar nuclear relationships between the evolution of several other orchids, their DNA diversification, and their nuclear ultrastructure. These findings give evidence for the suggestion that the complexity of the nuclear ultrastructure is indicative for the evolutionary state of a species.

We thank Mrs. S. KUHNER for technical assistance, and Dr. D. SCHWEIZER (Vienna) for critical reading of the manuscript.

References

BARLOW, P. W., 1977: Determinants of nuclear chromatin structure in angiosperms. — Ann. Sci. Nat., Bot. Biol. Vég. **18**, 193—206.
BENNETT, M. D., 1973: Nuclear characters in plants. — Brookhaven Symp. Biol. **25**, 344—366.
— SMITH, J. P., 1976: Nuclear DNA amounts in angiosperms. — Phil. Transact. R. Soc. London, B, **274**, 227—274.

COMINGS, D. E., OKADA, T. A., 1976: Fine structure of the heterochromatin of the kangoroo rat *Dipodomys ordii*, and examination of the possible role of actin and myosin in heterochromatin condensation. — J. Cell Sci. **21**, 465—477.

DAYAL, N., 1975: Genotypic control of chromocenters in radish (*Raphanus sativus* L. var. *radicola* PERSOON). — Caryologia **28**, 429—435.

DOUTRELIGNE, J., 1939: Les divers "types" de structure nucléaire et de mitose somatique chez les Phanérogames. — Cellule **48**, 191—215.

EICHHORN, A., 1934: Types définis et types intermédiaires dans la mitose des végétaux. — Cytologia **5**, 253—268.

INGLE, J., TIMMIS, J. N., SINCLAIR, J., 1975: The relationship between satellite DNA, rRNA gene redundancy, and genome size. — Pl. Physiol. **55**, 496—501.

LAFONTAINE, J. G., 1974: Ultrastructural organization of plant cell nuclei. — In BUSCH, H., (Ed.): The Cell Nucleus, vol. **1**, pp. 149—185. — New York: Academic Press.

NAGL, W., 1976 a: Early embryogenesis in *Tropaeolum majus* L.: Evolution of DNA content and polyteny in the suspensor. — Pl. Sci. Lett. **7**, 1—8.

— 1976 b: Zellkern und Zellzyklen. — Stuttgart: Ulmer.

— 1977: Nuclear structures during cell cycles. — In ROST, T. L., GIFFORD, E. M., JR. (Eds.): Mechanisms and Control of Cell Division, pp. 147—193. — Stroudsburg, PA.: Dowden, Hutchinson & Ross.

— 1978: Endopolyploidy and Polyteny in Differentiation and Evolution. — Amsterdam: North-Holland.

— 1979: Condensed chromatin in plant and animal cell nuclei: fundamental differences. — Pl. Syst. Evol., Suppl. **2**, 247—260.

— CAPESIUS, I., 1976: Molecular and cytological characteristics of nuclear DNA and chromatin for angiosperm systematics: Basic data for *Helianthus annus* (*Asteraceae*). — Pl. Syst. Evol. **126**, 221—237.

— 1977: Repetitive DNA and heterochromatin as factors of karyotype evolution in phylogeny and ontogeny of orchids. — Chromosomes Today **6**, 141—152.

— HABERMANN, T., FUSENIG, H.-P., 1977: Nuclear DNA contents in four primitive angiosperms. — Pl. Syst. Evol. **127**, 103—105.

OLSZEWSKA, M., 1973: Ultrastructure of the elementary fibril in extended and condensed chromatin and in metaphase chromosomes. — Acta Soc. Bot. Polon. **42**, 265—279.

SPURR, A. R., 1969: A low viscosity epoxy resin embedding medium for electron microscopy. — J. Ultrastr. Res. **26**, 31—43.

TANAKA, R., 1971: Types of resting nuclei in *Orchidoceae*. — Bot. Mag. Tokyo **84**, 118—122.

TSCHERMAK-WOESS, E., 1963: Strukturtypen der Ruhekerne von Pflanzen und Tieren. — Wien: Springer.

Address of the authors: Prof. Dr. WALTER NAGL, HANS-PETER FUSENIG, Department of Biology, The University, P.O. Box 3049, D-6750 Kaiserslautern, Federal Republic of Germany.

.

Pl. Syst. Evol., Suppl. 2, 235—245 (1979)
© by Springer-Verlag 1979

Department of Biology, The University of
Kaiserslautern, Federal Republic of Germany

Chromatin Organization and
Repetitive DNA in *Anacyclus* and *Anthemis (Asteraceae)*

By

Bärbel Fuhrmann and **Walter Nagl**, Kaiserslautern

Key Words: Angiosperms, *Asteraceae-Anthemideae*, *Anacyclus*, *Anthemis.*—Chromatin organization, nuclear ultrastructure, repetitive DNA.

Abstract: Genera and species of *Anthemideae* can be distinguished on the basis of their nuclear ultrastructure (pattern of chromatin condensation, amount of heterochromatin) and their DNA composition (proportion of repetitive DNA). *Anthemis* species exhibit chromomeric to chromonematic nuclei with some small, but well distinguishable chromocenters, *Anacyclus* species show highly chromonematic nuclei with a cap-like distribution of the chromonemata; heterochromatin cannot be distinguished in ultrathin sections of *Anacyclus* nuclei. The melting point of nuclear DNA is similar in all species studied (about 85 °C, corresponding to about 38% GC). The fractions of repetitive DNA classes are different in all species, indicating a rapid diversification of the genome composition in *Anthemideae*. There are rather complex relationships between the species-specific degree of chromatin condensation, amount of heterochromatin, nuclear DNA content (2 C value), and proportion of highly and intermediately repetitive DNA sequences.—It is suggested that nuclear ultrastructure and DNA composition may be used to characterize species and genera in sophisticated systematic and evolutionary studies.

Recently, studies on the genome and chromatin organization became a tool well suited to elaborate evolutionary relationships in the plant and animal kingdoms (e.g. CULLIS & SCHWEIZER 1974, MIZUNO & al. 1976, NARAYAN & REES 1976, BACHMANN & PRICE 1977, FLAVELI & al. 1977, NAGL & CAPESIUS 1977, CAPESIUS & NAGL 1978; see also reviews by NAGL 1978, 1979).

In this paper we report genome and chromatin characters from some *Anthemideae* species. This group is well studied in respect to karyotypes,

2 C values, developmental biology, phytochemistry, and systematics (e.g. EHRENDORFER 1959, 1964, MITSUOKA & EHRENDORFER 1972, NAGL 1974, NAGL & EHRENDORFER 1974, SCHWEIZER & EHRENDORFER 1976, EHRENDORFER & al. 1977, GREGER 1977, HEYWOOD & HUMPHRIES 1977). It is emphasized that nuclear ultrastructure and DNA composition are valuable markers for the evolution of species and genera, and that the complex relationships between nuclear parameters require caution in the generalization of correlations found within a certain genus.

Material and Methods

The material used is listed in Table 1. Seeds were obtained from Professor F. EHRENDORFER, Vienna, and the plants were grown in the Botanical Garden of the University of Kaiserslautern. Vouchers are preserved in the herbaria of both universities, Vienna and Kaiserslautern.

The electron microscopic and biochemical techniques are described in other publications (NAGL & FUSENIG 1979, SCHAAN & NAGL 1979, NAGL, in preparation).

Species-Specific Chromatin Structures

The nuclear ultrastructure of the two genera studied, *Anthemis* and *Anacyclus*, is clearly different. *Anthemis* nuclei belong to the chromomeric and chromonematic type, respectively, while *Anacyclus* nuclei display a cap-like distribution of the chromatin, i.e. a high concentration of chromatin at one pole of the nucleus (Figs. 1–3). While in *Anthemis* nuclei the heterochromatin can be distinguished from the euchromatin in electron micrographs by means of the different size of euchromatic and heterochromatic elements, this is impossible in *Anacyclus* nuclei. Here the euchromatin forms small patches (chromomeres) at the one pole of the nucleus, and a dense network of strands and clumps at the other pole, which hide the heterochromatin. The average proportion of the nuclear sections which is occupied by recognizable (i.e. condensed) chromatin increases with increasing 2 C value supporting the hypothesis that the chromatin texture of plant nuclei is nucleotypically controlled (NAGL & FUSENIG 1979). Table 1 shows more detailed data on the chromatin patterns. The species-specificity of the nuclear structure is clearly visible from these data. *Anthemis maritima*, the *Anthemis* species with the lowest DNA content, which was studied by means of electron microscopy (W. NAGL, unpublished), has nuclei of the chromomeric type plus some chromocenters; only 51 % of the nuclear area are occupied by chromatin. *Anthemis montana* has also chromomeric nuclei, but larger chromocenters, and 65 % of the nuclear area are occupied by chromatin (Fig. 2). The relatively high DNA content may therefore be due to both polyploidy and the higher

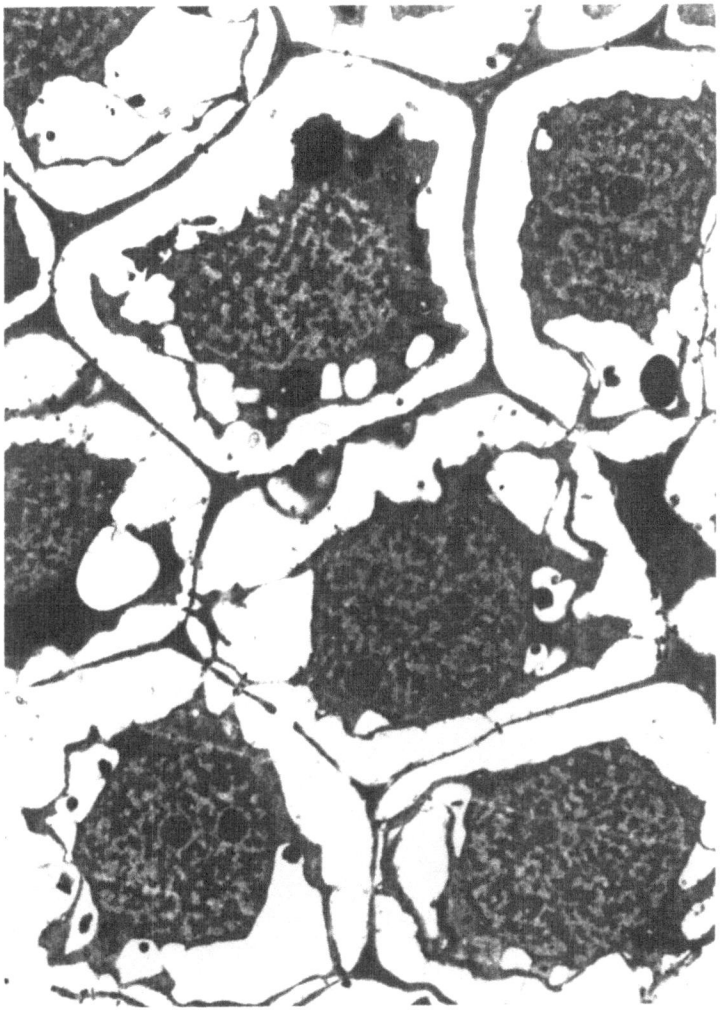

Fig. 1. Electron micrograph of a plasmolysed leaf meristem of *Anthemis altissima* (AN-743). Note the chromonematic organization of the nuclei. This survey figure was chosen to demonstrate that the chromonematic nature of the nuclei is not characteristic for a single cell cycle stage only (× 4,000)

heterochromatin content. *Anthemis altissima*, on the contrary, has only very little heterochromatin, so that the higher 2 C value is due to an increased euchromatin content. This is also expressed by the chromonematic structure of the nuclei (Fig. 1), and the 71 % chromatin occupying the nuclear area. Quantitative analysis of electron micrographs of *Anacyclus* nuclei is possible to some extent only, because euchromatin and heterochromatin are indistinguishable, and because the proportion of chromatin seen in the nucleus strongly depends on the site and direction of the section with respect to the cap-like chromatin distribution. Fig. 3 shows a nucleus sectioned "longitudinally", so that the cap-like organization is well visible. Sections through the cap or the opposite region give extremely different pictures.

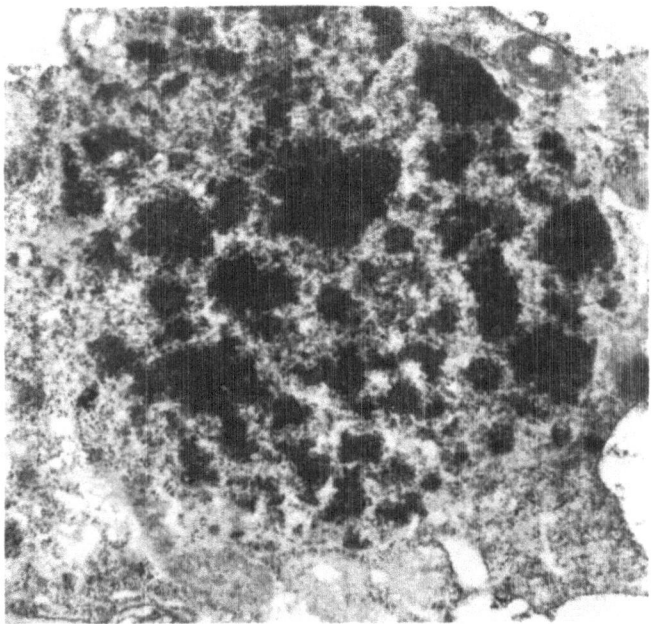

Fig. 2. Electron micrograph of an interphase nucleus of *Anthemis montana* (AN-763). Note the characteristic chromomeric organization of the euchromatin, and the chromocentric organization of the heterochromatin (× 16,800)

Species-Specific DNA Composition

DNA renaturation kinetics of the four species studied are shown in Fig. 4. In general, *Anacyclus* species possess more highly reiterated DNA than *Anthemis* species, but the size of each individual cot fraction is different in all species. As the melting point of the DNA's of all four species is very similar, about 85 % (Table 1), corresponding to an average base composition of 38 % GC, the variation of the percentage of repetitive DNA cannot be due to the extra synthesis (saltatory replication) or loss of a specific DNA sequence, but rather due to gain or

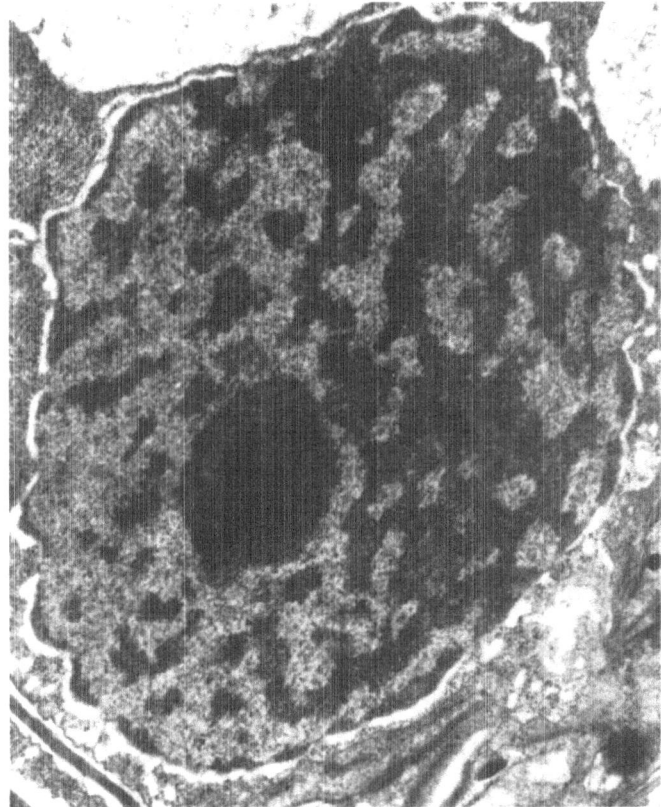

Fig. 3. Electron micrograph of a nucleus of *Anacyclus depressus* (AC-136). The chromatin shows a cap-like distribution, and euchromatin and heterochromatin cannot be distinguished. A nucleolus is visible in the center (× 16,800)

loss of DNA sequences scattered throughout the genome. Only 11 to 37 % of the genome correspond to single-copy sequences.

The redundancy (frequency) of the highest reiterated DNA sequences was calculated according to the formulae given by BRITTEN & al. (1974) and using *E. coli* DNA as a standard. In *Anthemis*, the fractions of highly repetitive DNA is reiterated 50,000 to 500,000 times, while in *Anacyclus* this fraction is reiterated 2 to 6×10^6 times (the precise data will be published elsewhere after the experiments have been repeated with another technique). The values obtained are comparable with that reported for another angiosperm, *Matthiola* (WENZEL 1979),

Table 1. Quantitative data on genome and

Species	2 C value (pg)	Haploid genome (pg)	T_m (°C)	Hyper-chrom. (%)	Unique DNA (%)	Repet. DNA (%)	Highly repet. DNA (%)
Anthemis montana L. (AN-763, 2n = 36)	16.9	4.2	85.6	39.2	37	63	14
Anthemis altissima L. (AN-743, 2n = 18)	15.8	7.9	85.3	39.9	33	67	18
Anacyclus depressus BALL (AC-136, 2n = 18)	12.4	6.2	84.3	37.4	11	89	44
Anacyclus radiatus LOIS. (AC-156, 2n = 18)	16.9	8.5	85.5	37.8	30	70	33

Nuclear DNA contents were measured cytophotometrically with a scanning photometer on Feulgen stained nuclei. DNA isolation, denaturation and renaturation is described in SCHAAN & NAGL (1979). As "highly repetitive DNA fraction" the DNA was taken which anneals faster than cot 10^{-2}.

although they are rather high for plants in general. In spite of those high degrees of redundancy, satellite DNA could not yet be detected by neutral CsCl gradient centrifugation.

Nucleotypic Relationships

Nucleotypic relationships concern 2 C values, proportions of repetitive DNA's, degree of chromatin condensation, heterochromatin content etc. (see also the report on *Anthemideae* by NAGL & EHRENDORFER 1974, NAGL 1974).

Fig. 5 d suggests that there is a weakly positive correlation between the percentage of reiterated DNA and the percentage of heterochromatin in *Anthemis*. Unfortunately, the heterochromatin of only two species was determined by means of electron microscopy. With respect to the amount of heterochromatin as visible after light microscopic staining techniques, a negative relationship between the percentage of total repetitive DNA and heterochromatin was found (Fig. 5 c). Therefore, the *Anthemideae* may represent another group of plants, in which heterochromatin is lacking considerable amounts of repetitive chromatin organization in some *Anthemideae*

Proportion occupying the nuclear area in electron micrographs (%)				Light microscopy: Heterochromatin (%) (approx.)
Chromatin (all visible)	Heterochromatin (% of chromatin)	Karyolymph	Nucleolus	
65	16	28	7	20.5
71	18	25	4	4.0
61	?	32	7	6.2
54	?	36	10	9.8

The percentage of nuclear substructures was calculated from gravimetric data from enlarged electron micrographs of 10 nuclei of each species (see also NAGL & FUSENIG 1979).

DNA, particularly highly repetitive sequences (Table 1; for *Tropaeolum* see DEUMLING & NAGL, 1978, for a general discussion see NAGL 1979). There is also a negative correlation between the percentage of repetitive DNA and 2 C nuclear DNA content (Fig. 5 a). This is in sharp contrast to the general rule that the variation in the 2 C content is mainly due to variation in the repetitive DNA (reviewed by NAGL 1968).

The percentage of chromatin in the electron-dense state when compared with the total nuclear area does not reveal any relationship to the amount of repetitive DNA, and only a slightly positive correlation with the 2 C content (Figs. 5 b). This can be either due to the small number of species studied, or because there is actually a deviation from the rule found in other taxa (NAGL 1979). In any case, these results show that we are still far away from an understanding of the determinants of nuclear ultrastructure and chromatin organization.

Conclusions

Comparative analysis of the nuclear ultrastructure in various *Anthemideae* species reveale fundamental differences in gross organization of their chromatin. There are types of organization which are characteristic for a genus, and there are species-specific variations of these types. The ultrastructure and texture of the nuclei can, therefore, be considered as a valuable characteristic for systematic studies. The

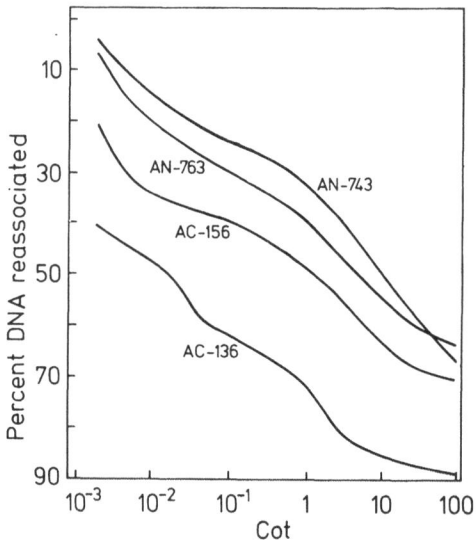

Fig. 4. Preliminary results of DNA reassociation kinetics (cot curves) of the four Anthemidean species studied: AN-763 = *Anthemis montana* var. *saxatilis*, AN-743 = *Anthemis   altissima*,   AC-156 = *Anacyclus   radiatus*,   AC-136 = *Anacyclus depressus*. Cot = Product of the DNA concentration at the beginning of the experiment (E_{260} of the native DNA), and the duration of the renaturation in seconds. Highly repetitive DNA was defined as DNA reassociated at a cot of 10^{-2}, unique DNA sequences were calculated to reassociate not before a cot between 10 and 100. The curves represent the means of 3 to 6 repeats

evolutionary relationships of the various types of nuclear organization are still difficult to analyse, because studies on nuclear ultrastructure exist only for a few plants, and also because its determinants are still incompletely understood. An approach to interpret nuclear structure in evolutionary terms has been made in orchids (TANAKA 1971, NAGL & CAPESIUS 1977, CAPESIUS & NAGL 1978) and ciliates (RAIKOV 1976). In the following paper (NAGL 1979) some possible trends in higher plants will be discussed.

The rule of a positive correlation between 2 C value and proportion or amount of repetitive DNA (reviewed by NAGL 1978) does not hold for the *Anthemideae*. These results are consistent with the assumption of different strategies in genome evolution (see also CULLIS & SCHWEIZER 1974, BALDARI & AMALDI 1976). Nevertheless, both the DNA content and

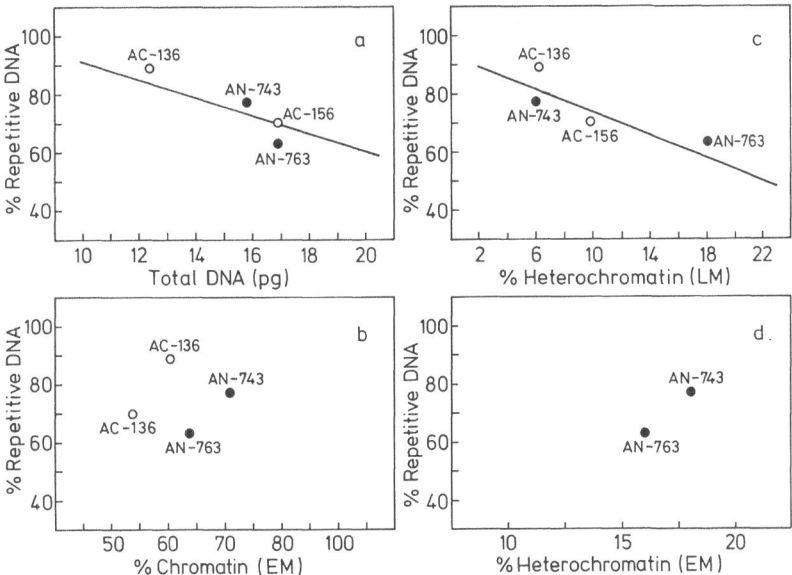

Fig. 5. Diagrams to illustrate the mainly negative relationships between proportion of repetitive DNA and some other nuclear characteristics: *a* total nuclear DNA content (2 C values; the relationship is only comparable, if the DNA content of the tetraploid species is halfed), *b* proportion of nuclear area is occupied by electron-dense chromatin in electron micrographs (no clear correlation can be seen in this diagram), *c* and *d* percentage of heterochromatin as visible in light and electron micrographs, respectively. The meaning of the codes is given in the legend to Fig. 4

the shape of cot curves (renaturation kinetics) may be valuable parameters in evolutionary studies in a tribe.

The determinants of nuclear fine structure, the degree of euchromatin condensation, the amount of heterochromatin etc. are still not very clear. It is very likely that there exist complex relationships between various repetitive DNA fractions and the texture of the nucleus.

The *Anthemideae* may serve as a valuable tool in molecular and cell biology as well as in systematic and evolutionary studies, because they

244   BÄRBEL FUHRMANN and W. NAGL:

apparently show evolutionary strategies which are different from that found in many other model systems. In addition, the *Anthemideae* may be an interesting system for developmental investigations, as their specific evolutionary strategies include uncommon developmental regulation mechanisms (NAGL & EHRENDORFER 1974, NAGL 1974).

We are thankful to Mrs. S. KÜHNER and Miss B. HOLZHAUSER for technical assistance, and Miss D. FRÖLICH for some data of *Anthemis maritima* DNA. Support by grant Na-107/3 of the *Deutsche Forschungsgemeinschaft* is gratefully acknowledged.

References

BACHMANN, K., PRICE, H. J., 1977: Repetitive DNA in *Cichorieae* (*Compositae*). — Chromosoma **61**, 267—275.

BALDARI, C. T., AMALDI, F., 1976: DNA reassociation kinetics in relation to genome size in four amphibian species. — Chromosoma **59**, 13—22.

BRITTEN, R. J., GRAHAM, D. E., NEUFELD, B. R., 1974: Analysis of repeating DNA sequences by reassociation. — Meth. Enzymol. **29 E**, 363—406.

CAPESIUS, I., NAGL, W., 1978: Molecular and cytological characteristics of nuclear DNA and chromatin for angiosperm systematics: DNA diversification in the evolution of four orchids. — Pl. Syst. Evol. **129**, 143—166.

CULLIS, C. A., SCHWEIZER, D., 1974: Repetition DNA in some *Anemone* species. — Chromosoma **44**, 417—421.

DEUMLING, B., NAGL, W., 1978: DNA characterization, satellite DNA localization, and nuclear organization in *Tropaeolum majus*. — Cytobiologie **16**, 412—420.

EHRENDORFER, F., 1959: Differentiation-hybridization cycles and polyploidy in *Achillea*. — Cold Spring Harbor Symp. Quant. Biol. **24**, 141—152.

— 1964: Notizen zur Cytotaxonomie und Evolution der Gattung *Artemisia*. — Österr. Bot. Z. **111**, 84—142.

— SCHWEITZER, D., GREGER, H., HUMPHRIES, C., 1977: Chromosome banding and synthetic systematics in *Anacyclus* (*Asteraceae-Anthemideae*). — Taxon **26**, 387—394.

FLAVELL, R. B., RIMPAU, J., SMITH, D. B., 1977: Repeated sequence DNA relationships in four cereal genomes. — Chromosoma **63**, 205—222.

GREGER, H., 1977: *Anthemideae* — chemical review. — In HEYWOOD, V. H., HARBORNE, J. B., TURNER, B. L., (Eds.): The Biology and Chemistry of the *Compositae*, vol. **2**, pp. 899—941. — London: Academic Press.

HEYWOOD, V. H., HUMPHRIES, C. J., 1977: *Anthemideae* — systematic review. In HEYWOOD, V. H., HARBORNE, J. B., TURNER, B. L., (Eds.): The Biology and Chemistry of the *Compositae*, vol. **2**, pp. 851—898. — London: Academic Press.

MITSUOKA, S., EHRENDORFER, F., 1972: Cytogenetics and evolution of *Matricaria* and related genera (*Asteraceae-Anthemideae*). — Österr. Bot. Z. **120**, 155—200.

MIZUNO, S., ANDREWS, C., MACGREGOR, H. C., 1976: Interspecific "common" repetitive DNA sequences in salamanders of the genus *Plethodon*. — Chromosoma **58**, 1—31.

NAGL, W., 1974: Mitotic cycle time in perennial and annual plants with various amounts of DNA and heterochromatin. — Devel. Biol. **39**, 342—346.

NAGL, W., 1978: Endopolyploid and Polyteny in Differentiation and Evolution. — Amsterdam: North-Holland.
— 1979: Condensed interphase chromatin in plant and animal cell nuclei: fundamental differences. — Pl. Syst. Evol., Suppl. **2**, 247—260.
— in preparation: Repetitive DNA in primitive and progressive angiosperms.
— CAPESIUS, I., 1977: Repetitive DNA and heterochromatin as factors of karyotype evolution in the phylogeny and ontogeny of orchids. — Chrom. Today **6**, 141—150.
— EHRENDORFER, F., 1974: DNA content, heterochromatin, mitotic index, and growth in perennial and annual *Anthemideae (Asteraceae)*. — Pl. Syst. Evol. **123**, 35—54.
— FUSENIG, H.-P., 1979: Types of chromatin organization in plant nuclei. — Pl. Syst. Evol., Suppl. **2**, 221—233.
NARAYAN, R. K. J., REES, H., 1976: Nuclear DNA variation in *Lathyrus*. — Chromosoma **54**, 141—154.
RAIKOV, I. B., 1976: Evolution of macronuclear organization. — Ann. Rev. Genet. **10**, 413—440.
SCHAAN, M., NAGL, W., 1979: Repetitive DNA in primitive angiosperms. — Pl. Syst. Evol., Suppl. **2**, 67—71.
SCHWEIZER, D., EHRENDORFER, F., 1976: Giemsa banded karyotypes, systematics, and evolution in *Anacyclus (Asteraceae-Anthemideae)*. — Pl. Syst. Evol. **126**, 107—148.
TANAKA, R., 1971: Types of resting nuclei in *Orchidaceae*. — Bot. Mag. Tokyo **84**, 118—122.
WENZEL, W., HEMLEBEN, V., 1979: DNA reassociation studies and considerations on the genome organization and evolution of higher plants. — Pl. Syst. Evol., Suppl. **2**, 29—40.

Address of the authors: Dipl.-Biol. BÄRBEL FUHRMANN, Prof. Dr. WALTER NAGL, Department of Biology, The University, P.O. Box 3049, D-6750 Kaiserslautern, Federal Republic of Germany.

Pl. Syst. Evol., Suppl. 2, 247—260 (1979)

Department of Biology, The University
of Kaiserslautern, Federal Republic of Germany

Condensed Interphase Chromatin in Plant and Animal Cell Nuclei: Fundamental Differences

By

Walter Nagl, Kaiserslautern

Key Words: Chromatin condensation, heterochromatin, DNA differential replication, RNA synthesis.

Abstract: In electron micrographs of interphase nuclei of both plants and animals, electron-dense chromatin can be found in variable amounts. Although this condensed chromatin looks alike in all micrographs, it covers at least *four different* classes of chromatin: constitutive heterochromatin (which can be visualized in plants and animals e.g. by the Giemsa banding technique), facultative heterochromatin (female sex chromatin in mammals), inactivated euchromatin (in animals only), and species-specific condensed euchromatin (in plants only). Care-free interpretation of condensed chromatin as *heterochromatin* may cause, therefore, much confusion. The possible molecular mechanisms involved in the condensation of the various classes of chromatin, and the suitability of condensed chromatin as marker for differential DNA and RNA synthesis, respectively, are discussed.

Some of the most frustrating aspects of cell biology are the difficulties to distinguish between euchromatin and heterochromatin in electron micrographs (unfortunately many workers are not aware of this!), the difficulty to analyze the molecular mechanisms of chromatin condensation (because there are evidently more than one), and to interpret them in terms of template activity. Where is the delimitation between chromomeres and chromocenters, if it is just a question of size, whether a band of a giant chromosomes reacts as euchromatin or as heterochromatin after differential Giemsa staining? (see HÄGELE 1977). What is the difference between euchromatin and heterochromatin in plants, when both exhibit the same electron-density in many species, and when both fail to hybridize with highly repetitive DNA? (see NAGL 1977 a, DEUMLING & NAGL 1978, FUHRMANN & NAGL 1979).

Nuclear structure and ultrastructure have been reported to be specific for certain species and genera (resulting in the elaboration of classification systems, e.g. DELAY 1948, TSCHERMAK-WOESS 1963, TANAKA 1971, LAFONTAINE 1974, BARLOW 1977, NAGL & FUSENIG 1979), specific for certain tissues, cells, and developmental stages (TSCHERMAK-WOESS 1963, NAGL 1976, JEANNY & GONTCHAROFF 1978, KUNZE & al. 1978), and specific for the cell cycle stage (reviewed by NAGL 1977 b). This report deals with a comparison of species-specific (function-independent) and tissue-specific (function-dependent) chromatin ultrastructures in plants and animals, respectively. The findings call for some caution in the interpretation of electron micrographs in terms of the molecular mechanism and functional significance of chromatin condensation. Some fundamental differences between plant and animal nuclei are revealed, so that their condensed chromatin can be used as marker for quite different aspects of cell metabolism.

Condensed Chromatin in Plants

As shown in a previous paper NAGL & FUSENIG 1979), plant cell nuclei display species-specific ultrastructures. In principle, there are two kinds of chromatin patterns. In the one, the euchromatin is diffuse or decondensed, and the heterochromatin forms electron-dense patches, which usually are called chromocenters (Fig. 1 b). In the other type of patterns both the euchromatin and the heterochromatin are condensed into electron-dense structures. In this case the euchromatin forms chromomeres and chromonemata, respectively, and the heterochromatin forms chromocenters, which can be distinguished from the euchromatic structures only if they are clearly larger than the latter. Fig. 1 a shows a chromonematic nucleus, which is completely euchromatic in spite of the obviously dense chromatin that can be seen. Therefore, the interpretation of electron micrographs of plant nuclei requires the comparative study of nuclear structure at the light microscope level.

Another aspect of nuclear ultrastructure in plants can be seen in the fact that functional condensation and decondensation of chromatin as a control mechanism of large-scale gene inactivation and activation, respectively, does not occur. Exceptions may be found in polytene nuclei only (NAGL 1969, 1974). The constancy of the species-specific degree of chromatin condensation allows to interpret changes in the proportion of heterochromatin in terms of differential heterochromatin replication, if 1. the heterochromatin can be clearly distinguished from the euchromatin, and 2. the direction of the change (extra- *versus* under-replication) can be deduced from the situation in diploid, meristematic

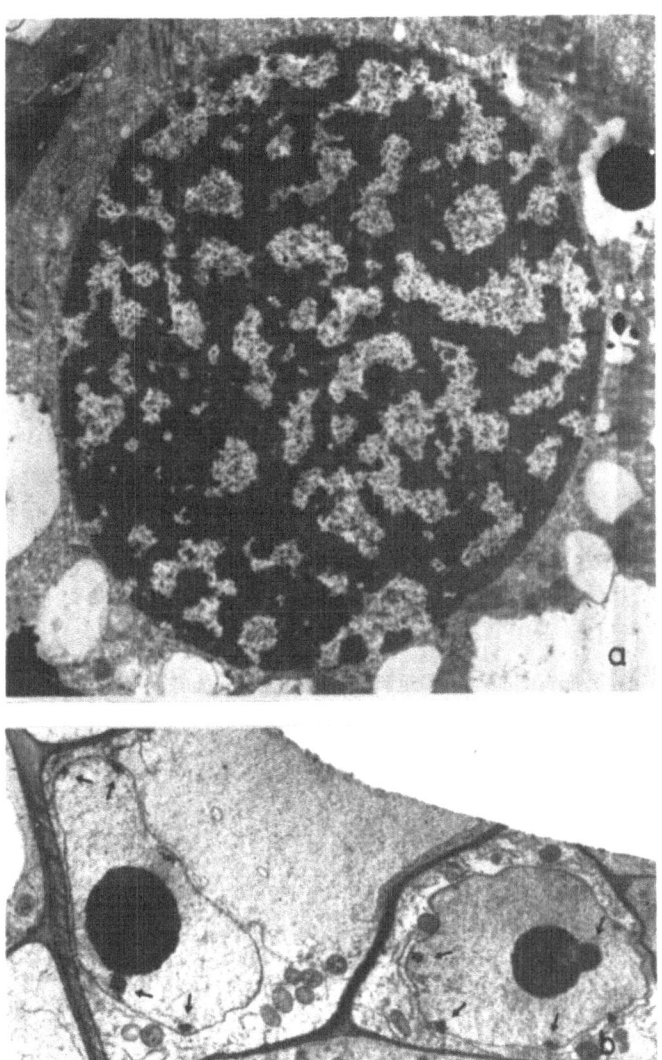

Fig. 1. Chromatin organization in the plant with the highest and lowest 2C value, respectively (4,240 ×). *a Viscum album* L. s.l. the euchromatin forms an electron-dense chromonema (reticulum), heterochromatin cannot be differentiated. *b Arabidopsis thaliana*, (L.) HEYNH. the euchromatin is completely decondensed, the heterochromatin forms some small chromocenters (arrows).
The large dark area in the center of the nuclei is the nucleolus

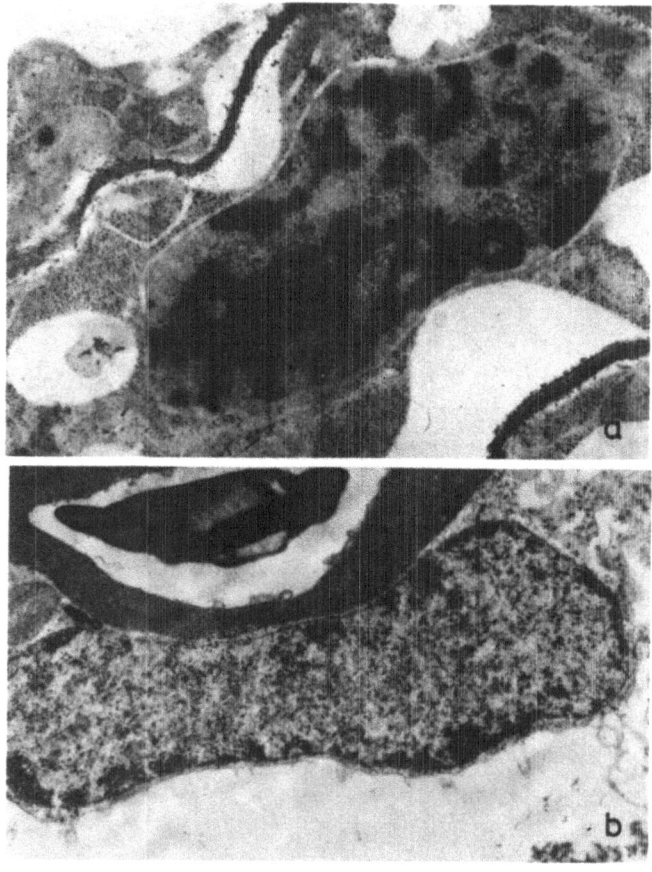

Fig. 2. Nuclear ultrastructure in juvenile and adult leaves of the ivy, *Hedera helix* L. (16,000 ×). *a* Nucleus from the juvenile phase with decondensed euchromatin and large masses of heterochromatin. *b* Nucleus from the adult phase with a lower proportion of heterochromatin (tetraploid nucleus which has undergone heterochromatin underreplication)

nuclei, or can be supported by DNA measurements. Fig. 2 gives an example of the morphology of heterochromatin underreplication in the ivy, *Hedera helix*. Fig. 2 *a* shows a meristematic nucleus of the species-specific structure (high proportion of constitutive heterochromatin). Fig. 2 *b* shows a tetraploid nucleus from the adult phase with much less heterochromatin. Scanning cytophotometric and biochemical DNA analyses have proven differential DNA replication in this species (KESSLER & RECHES 1977, SCHÄFFNER & NAGL 1979). Other examples were reviewed by NAGL (1978).

The opposite case, heterochromatin amplification, was found in organ cultures of the orchid *Cymbidium* (NAGL 1972). On grounds of the heterochromatin content of diploid, mitotically active cells it could be ensured by cytophotometry and autoradiography that there is actually a disproportionate *increase* of heterochromatin during polyploidization, and not only a change in the proportion of the condensation state of euchromatin. Further studies employing Giemsa and fluorochrome

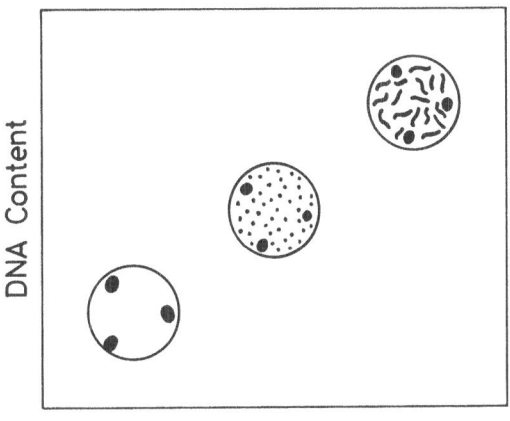

Fig. 3. Diagram to illustrate the relationship between the basic nuclear DNA content (2 C value) and the compactness of euchromatin (species-specific nuclear structure) in plants. Heterochromatin is shown in the form of chromocenters, but heterochromatin may be absent in chromomeric and chromonematic nuclei

staining (SCHWEIZER & NAGL 1976) and biochemical DNA analysis (NAGL & RÜCKER 1976) have supported this interpretation.

Summarizing it can be stated that variation in the amount of condensed chromatin *between* plant species is only in part due to variation in heterochromatin, but largely due to variation in the degree of condensation of euchromatin. As will be discussed later this degree of condensation is determined mainly by the DNA content (2 C value) and DNA composition (Fig. 3). The variation in the amount of condensed chromatin *within* a plant, on the other hand, is the expression of differential heterochromatin replication, because no functional condensation and decondensation mechanisms are known to occur in higher plants (Fig. 4). Exceptions from this rule may only occur in highly specialized polytenic nuclei (NAGL 1974). This is, however, rather a

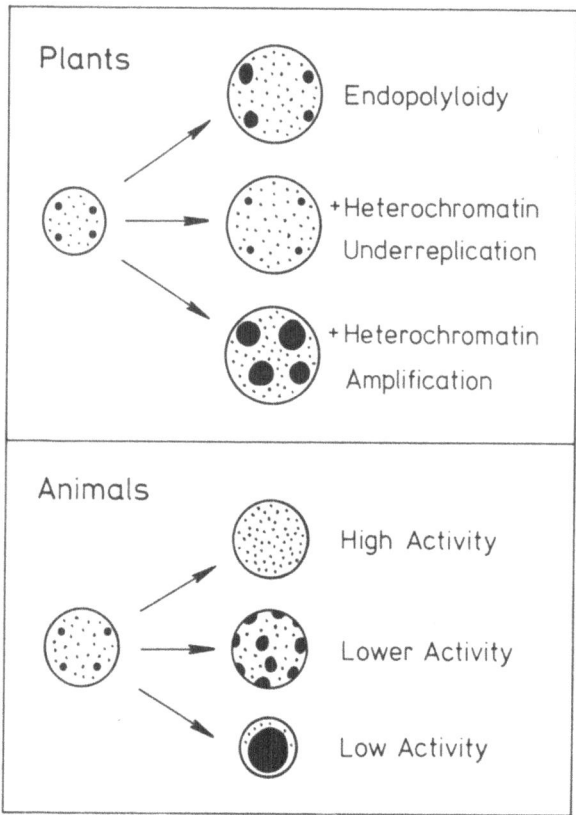

Fig. 4. Diagrams to illustrate, how to interpret the variation of nuclear structure within an individual organism in plants and animals. In plants, all heterochromatin that is visible represents constitutive heterochromatin. Variation in heterochromatin during somatogenesis is the expression of total or differential heterochromatin replication. In animals most of the condensed chromatin represents euchromatin. Variation of condensed euchromatin is the expression of variation in template activity

question of assembling or separation of the multiple sister chromosomes (endo-chromosomes) than of condensation or decondensation of the individual chromatid fiber.

Fig. 5. Chromatin organization in various cell types of the Guinea-pig (12,800 ×). a Cornea: nearly all chromatin is decondensed. b Pigmented epithelium of the eye: the nucleus exhibits several masses of condensed chromatin. c A macrophage nucleus with predominately condensed chromatin. As all three nuclei have the same diploid DNA content, the various structures must be the result of different degrees of euchromatin condensation

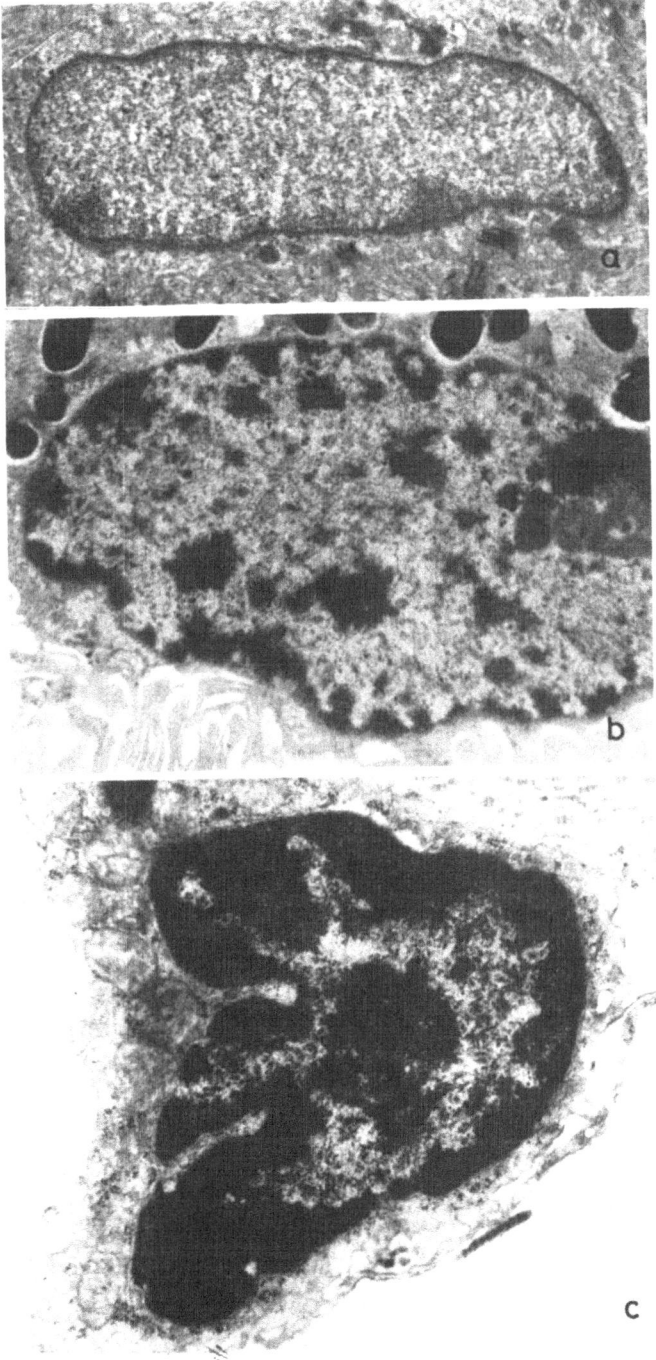

Figs. 5 *a–c*

Condensed Chromatin in Mammals

Nuclei of animals, particularly of mammals, often display cell- and tissue-specific ultrastructures so that no species-specific chromatin organization can be detected. While mitotically active cells often exhibit completely "euchromatic" nuclei, i.e. all the chromatin fractions are in the decondensed state, mature tissues show nuclei with various degrees of chromatin condensation (reviewed by TSCHERMAK-WOESS 1963; see also PERA & WOLF 1967). Fig. 5 shows nuclei from various tissues of the Guinea pig, *Cavia cobaya*, with different degrees of chromatin condensation, but the same nuclear DNA content as estimated by Feulgencytophotometry of the same cell types. As only a small portion of the condensed chromatin represents constitutive heterochromatin (i.e. Giemsa-positive, late replicating heterochromatin), that variation must be interpreted as expression of differential chromatin inactivation. The transcriptional inactivity of condensed chromatin has actually been proven in many cases (reviewed by NAGL 1976; see also the results in the Guinea pig retina, outlined by SCHMALENBERGER & NAGL 1979). Inactive nuclei are completely condensed (e.g. in hen erythrocytes, mammalian sperm cells). Therefore, variation in the amount of condensed chromatin in mammalian cells can be serve as indication of nuclear activity or inactivity (Fig. 4).

Summarizing it can be stated that no species-specific nuclear structure can be found in mammals (and many other animal groups). The enormous variation of nuclear structure, as expressed by variable degrees of chromatin condensation, is function-dependent and tissue-specific and cannot used, therefore, as an indication of differential DNA replication as in plants, but can be seen as a manifestation of differences in the transcription activity.

A Warning Against the Unqualified Use of the Term Heterochromatin

HEITZ (1928) defined heterochromatin as that portion of the chromatin that remains condensed throughout the cell cycle except during a short period of the prophase, the so-called *"Zerstäubungs-stadium"* (dispersion stage). Today the constitutive or karyotypical heterochromatin is characterized by a) its high content in simple-sequence repetitive DNA, b) its asynchronous (normally late) DNA replication when compared with the DNA in euchromatin, c) its frequent differential replication in endoreduplication cycles (normally under-replication), d) its inertness in structural genes and its low RNA synthesis, e) its position effect on neighboring genes, and f) its differential stainability with Giemsa.

More detailed reviews and examples are given by NAGL (1976 and 1978).

In addition to constitutive heterochromatin, which is one of the characteristics of the species-specific karyotype, facultative heterochromatin occurs in female mammals. This is also known as sex chromatin or BARR body, and represents one of the two X chromosomes that becomes inactivated and condensed during embryogenesis (LYON 1968). Although facultative heterochromatin is late replicating, it lacks many other aspects of constitutive heterochromatin.

In nuclei of mammals, however, often much more chromatin than facultative and constitutive heterochromatin is in a condensed state (Fig. 5). This *condensed euchromatin* does not show, of course, any similarity to heterochromatin except that it is condensed and transcriptionally inactive. This has caused much confusion because several authors have equated all condensed chromatin with heterochromatin (e.g. FRENSTER 1974). Although the old question of whether heterochromatin is a certain *species* of chromatin, or just a certain *state* of it, today cannot be answered beyond doubt, it seems misleading to me, and inappropriate, to identify condensed chromatin and constitutive heterochromatin. Particularly, the species-specific condensation of euchromatin in plants (which exhibits the same electron-density as heterochromatin), and the tissue-specific condensation (inactivation) of large portions of the chromatin in animals should not be equated with the permanent and karyotypical condensation of heterochromatic portions of the chromosomes. The use of the same term for all kinds of condensed chromatin easily could obscure fundamental differences in their molecular organization.

Mechanisms of Chromatin Condensation

Constitutive heterochromatin has been found to be often rich in highly repetitive DNA such as satellite DNA (e.g. ECKHARDT & GALL 1971, YUNIS & YASMINEH 1971, REES & al. 1976, RENKAWITZ 1978a; reviewed by NAGL 1976). Several mechanisms, therefore, have been discussed, by which highly repetitive DNA is involved in chromatin condensation. A very likely possibility is that satellite DNA preferentially binds to certain classes of histones, e.g. phosphorylated H 1, which cause chromatin condensation (BLUMENFELD & al. 1978). Constitutive heterochromatin can, however, also be deficient in repetitive DNA sequences (ARRIGHI & al. 1974), or it may lack detectable amounts of satellite DNA (CAPESIUS & NAGL 1978, DEUMLING & NAGL 1978, FUHRMANN & NAGL 1979). It rather seems that the various types of heterochromatin which are known to exist from differential staining techniques (GREILHUBER 1975, NAGL 1976), are the results of either a

certain sequence organization *or* the result of a high proportion, *or* a high frequency of repetitive DNA sequences (e.g.
Renkawitz 1978 b). But it cannot yet be excluded that just the *amount of*
DNA with recognition sites for certain proteins (histone H 1 ?)
determines the species-specific degree of chromatin condensation. The
idea that heterochromatin formation primarily depends on the amount
of DNA per chromosome unit is strongly supported by the finding of
Hägele (1977) that "euchromatic" bands in polytene chromosomes
become "heterochromatic" (Giemsa-positive) after duplication of that
segment, and by the increasing overall chromatin condensation in plants
with increasing 2 C DNA values (Fig. 3); Nagl 1976, 1977 a, Barlow
1977, Nagl & Fusenig 1979).

Condensation of euchromatin as a mechanism (and morphological
expression) of large-scale gene inactivation is understood as the result of
histone modification and the influence of non-histone chromosomal
proteins (see the Cold Spring Harbor Symp. Quant. Biol. 1974, 1978,
Nagl 1976). The highly variable structure of mammalian cell nuclei—
expressed, for instance, by decondensation of both heterochromatin and
euchromatin, or by differential states of condensation, or by condensation of both types of chromatin—indicates that chromosomal
proteins can affect the compactness of chromatin independently of its
DNA composition. It is now widely accepted that all chromatin consists
of repeated subunits, the nucleosomes, which are tandemly arranged like
beads on a string (Olins & Olins 1974, Oudet & al. 1975), and that
condensation of this chain to thicker chromatin fibrils and to condensed
chromatin is influenced by the extra-nucleosomal histone H 1 (e.g.
Sluyser 1977, Thoma & Koller 1977, Worcel & Benyajati 1977,
Hildebrand & al. 1978). Thus, it is evident that histones are a factor of
both heterochromatin and functionally condensed euchromatin. In
addition, H 1 modification has been detected to play some role in mitotic
chromosome condensation (e.g. Bradbury 1975, Gurley & al. 1978).
Therefore, there is a great need for studies of histones in plants and
animals, in order to understand the various aspects of condensed
chromatin.

Summarizing we can state that there is often a close relationship
between the amount of heterochromatin and highly reiterated DNA
sequences, but there are also contradicting findings, so that this
relationship must not be generalized. Similarly, there is often a positive
correlation between the total nuclear DNA content (2 C value) and the
amount of DNA sequences with intermediate reiteration (Narayan &
Rees 1976, 1977, Gaddipati 1976, Macgregor & al. 1976, reviewed by
Nagl 1976, 1978). But this correlation too does not hold for all taxa. So it
seems that gross organization of chromatin and nuclear ultrastructure

are determined in some way be the DNA mass and by the proportion of highly and intermediately reiterated DNA sequences (see also LERMAN & DEGTYAREV 1978), but that there exist rather complex relationships between DNA sequence organization, histone modification, euchromatin condensation, and amount of heterochromatin.

Euchromatin Compactness and Transcription

In the previous paragraphs it was shown that there is enormous variation in the amount and degree of chromatin condensation among species and tissues. During the last years multiple evidence has been obtained for a close relationship between decondensation of chromatin and increase in transcriptional activity (e.g. puffs; see also the preceding paragraphs). This relationship was, however, obtained nearly exclusively in animal cells. In plants, only endomitotic and polytenic nuclei behave similarly (NAGL 1969, 1973). Although it was found that decondensed chromatin incorporated more ^{3}H-uridine than condensed chromatin, the *species-specific variation* of chromatin condensation is much larger than the *tissue-specific* one. It is very likely that the transcriptional activity in a root tip cell of *Allium* and *Brassia* is similar, although all the chromatin is condensed into chromonemata in the former, while it is completely diffuse and decondensed in the latter. Experiments are now under way to measure the quantity of RNA synthesized in nuclei of different structural types, because caution is recommended in the interpretation of the degree of chromatin condensation in plants in terms of different activity.

Our understandig of the functional organization of chromatin is based on a few model systems only; but we still lack information on nuclear ultrastructure and nuclear activity for most species and cell types. Before we can generalize sporadic findings and formulate new principles of molecular and cell biology we need much more basic "descriptive" studies of nuclear structure and function.

Conclusions

1) All degrees of morphological transition occur between euchromatin and heterochromatin, e.g. condensed euchromatin, or decondensed heterochromatin.

2) The molecular mechanisms of euchromatin condensation may be different in plants and animals. In plants, euchromatin condensation is species-specific, function-independent, and probably determined by the DNA mass and genome organization. In animals (particularly mammals), euchromatin condensation is tissue-specific, function-dependent, and evidently controlled by the histone metabolism as a step in gene inactivation.

3) Heterochromatin, although exhibiting the same electron density as euchromatin, must not be equated with condensed euchromatin, because there may be differences with respect to the condensation process and the genetic and functional states. Highly repetitive DNA cannot be seen as an essential characteristic of heterochromatin.

4) Variation of condensed chromatin among nuclei of the same individual appears as a marker for differential DNA replication in plants, but as a marker for different transcriptional activity in animals.

I thank Mrs. S. KUHNER for careful technical assistance, and Mr. B. SCHMALENBERGER for some of the electron micrographs.

References

ARRIGHI, F. E., HSU, T. C., PATHAK, S., SAWADA, H., 1974: The sex chromosomes of the Chinese hamster: constitutive heterochromatin deficient in repetitive DNA sequences. — Cytogenet. Cell Genet. **13**, 268—274.

BARLOW, P. W., 1977: Determinants of nuclear chromatin structure in Angiosperms. — Ann. Sci. Nat., Bot. & Biol. Végét., Ser. 12, **18**, 193—206.

BLUMENFELD, M., ORF, J. W., SINA, B. J., KREBER, R. A., CALLAHAN, M. A., MULLINS, J. I., SNYDER, L. A., 1978: Correlation between phosphorylated H 1 histones and satellite DNAs in *Drosophila virilis*. — Proc. Nat. Acad. Sci. (U.S.) **75**, 866—870.

BRADBURY, E. M., 1975: Histones, chromatin structure, and control of cell division. — Current Topics Devel. Biol. **9**, 1—13.

CAPESIUS, I., NAGL, W., 1978: Molecular and cytological characteristics of nuclear DNA and chromatin for angiosperm systematics: DNA diversification in the evolution of four orchids. — Pl. Syst. Evol. **129**, 143—166.

Cold Spring Harbor Symp. Quant. Biol. 1974: The Eukaryotic Chromosome (vol. **38**).

— 1978: Chromatin (vol. **42**).

CULLIS, C. A., SCHWEIZER, D., 1974: Repetition DNA in some *Anemone* species. — Chromosoma **44**, 417—421.

DELAY, C., 1948: Recherches sur la structure des noyaux quiescents chez les Phanérogames. — Rev. Cytol. Cytophysiol. vég. **10**, 103—228.

DEUMLING, B., NAGL, W., 1978: DNA characterization, satellite DNA localization, and nuclear organization in *Tropaeolum majus*. — Cytobiologie **16**, 412—420.

ECKHARDT, R. A., GALL, J. G., 1971: Satellite DNA associated with heterochromatin in *Rhynchosciara*. — Chromosoma **32**, 407—427.

FRENSTER, J. H., 1974: Ultrastructure and function of heterochromatin and euchromatin. — In BUSCH, H., (Ed.): The Cell Nucleus, vol. **1**, pp. 565—580. — New York: Academic Press.

FUHRMANN, B., NAGL, W., 1979: Chromatin organization and repetitive DNA in *Anacyclus* and *Anthemis* (*Asteraceae*). — Pl. Syst. Evol., Suppl. **2**, 235—245.

GADDIPATI, J. P. B., 1976: Characterization of chromosomal DNA in higher plants with varying size. — Thesis, Dept. Genetics & Plant Breeding, Banaras Hindu University, India.

GREILHUBER, J., 1975: Heterogeneity of heterochromatin in plants: comparison of Hy- and C-bands in *Vicia faba*. — Pl. Syst. Evol. **124**, 139—156.

GURLEY, L. R., D'ANNA, J. A., BARHAM, S. S., DEAVEN, L. L., TOBEY, R. A., 1978: Histone phosphorylation and chromatin structure during mitosis in Chinese hamster cells. — Europ. J. Biochem. **84**, 1—15.

HÄGELE, K., 1977: Differential staining of polytene chromosome bands in *Chironomus* by Giemsa banding methods. — Chromosoma **59**, 207—216.

HEITZ, E., 1928: Heterochromatin der Moose, I. — Jahrb. Wiss. Bot. **69**, 762—818.

HILDEBRAND, C. E., TOBEY, R. A., GURLEY, L. R., WALTERS, R. A., 1978: Action of heparin on mammalian nuclei. II. Cell-cycle-specific changes in chromatin organization correlate temporally with histone H 1 phosphorylation. — Biochim. Biophys. Acta **517**, 486—499.

JEANNY, J. C., GONTCHAROFF, M., 1978: Etude en microscopie électronique et par cytophotométrie à balayage de la structure et de la distribution de la chromatine dans les noyaux des cellules cartilagineuses de *Triturus cristatus* âgés au cours de la régéneration. — W. Roux's Arch. **184**, 195—211.

JONES, K. W., 1976: Repetitive DNA sequences in animals, particularly primates. — Chrom. Today **5**, 305—313.

KESSLER, B., RECHES, S., 1977: Structural and functional changes of chromosomal DNA during aging and phase change in plants. — Chrom. Today **6**, 237—246.

KUNZE, K. D., HERRMANN, W. R., MEYER, W., 1978: An image analysing method for automated cytology prescreening of cervix carcinoma and its prestages: demonstration and preliminary results. — Arch. Geschwulstforsch. **48**, 131—139.

LAFONTAINE, J. G., 1974: Ultrastructural organization of plant cell nuclei. — In BUSCH, H., (Ed.): The Cell Nucleus, vol. **1**, pp. 149—185. — New York: Academic Press.

LERMAN, M. I., DEGTYAREV, S. V., 1978: Periodically interspersed repetitive sequences may govern higher order DNA coiling in chromatin and chromosomes. — Mol. Biol. Rep. **4**, 117—120.

LYON, M. F., 1968: Chromosomal and subchromosomal inactivation. — Ann. Rev. Genet. **2**, 31—52.

MACGREGOR, H. C., MIZUNO, S., VLAD, M., 1976: Chromosomes and DNA sequences in salamanders. — Chrom. Today **5**, 331—339.

NAGL, W., 1969: Correlation of structure and RNA synthesis in the nucleolus-organizing polytene chromosomes of *Phaseolus vulgaris*. — Chromosoma **28**, 85—91.

— 1972: Evidence of DNA amplification in the orchid *Cymbidium in vitro*. — Cytobios **5**, 145—154.

— 1973: The mitotic and endomitotic nuclear cycle in *Allium carinatum*. IV. ³H-Uridine incorporation. — Chromosoma **44**, 203—212.

— 1974: The *Phaseolus* suspensor and its polytene chromosomes. — Z. Pflanzenphysiol. **73**, 1—44.

— 1976: Zellkern und Zellzyklen.— Stuttgart: Ulmer.

— 1977 a: Early and late DNA replication in respectively condensed and decondensed heterochromatin of *Allium carinatum*. — Protoplasma **91**, 389—407.

— 1977 b: Nuclear structures during cell cycles. In ROST, T. L., GIFFORD, E. M., JR., (Eds.): Mechanisms and Control of Cell Division, pp. 147—193. — Stroudsburg, Pa., USA: Dowden, Hutchison & Ross.

— 1978: Endopolyploidy and Polyteny in Differentiation and Evolution. — Amsterdam: North-Holland.

17*

NAGL, W., FUSENIG, H.-P., 1979: Types of chromatin organization in plant cell nuclei. — Pl. Syst. Evol., Suppl. 2, 221—233.

NAGL, W., RÜCKER, W., 1976: Effects of phytohormones on thermal denaturation profiles of *Cymbidium* DNA: Indication of differential DNA replication. — Nucl. Acids Res. 3, 2033—2039.

NARAYAN, R. K. J., REES, H., 1976: Nuclear DNA variation in *Lathyrus*. — Chromosoma 54, 141—154.

— — 1977: Nuclear DNA divergence among *Lathyrus* species. — Chromosoma 63, 101—107.

OLINS, A. L., OLINS, D. E., 1974: Spheroid chromatin units (v bodies). — Science 183, 330—332.

OUDET, P., GROSS-BELLARD, M., CHAMBON, B., 1975: Electron microscopic and biochemical evidence that chromatin structure is a repeating unit. — Cell 4, 281—300.

PERA, F., WOLF, U., 1967: DNS-Replikation und Morphologie der X-Chromosomen während der Syntheseperiode bei *Microtus agrestis*. — Chromosoma 22, 378—389.

REES, R. W., FOX, D. P., MAHER, E. P., 1976: DNA content, reiteration and satellites in *Dermestes*. — In JONES, K., BRANDHAM, P. E., (Eds.): Current Chromosome Research, pp. 33—41. — Amsterdam: North-Holland.

RENKAWITZ, R., 1978 a: Two highly repetitive DNA satellites of *Drosophila hydei* localized within the α-heterochromatin. of specific chromosomes. — Chromosoma 66, 237—248.

— 1978 b: Characterization of two moderately repetitive DNA components localized within the β-heterochromatin of *Drosophila hydei*. — Chromosoma 66, 225—236.

SCHÄFFNER, K.-H., NAGL, W., 1979: Differential DNA replication involved in transition from juvenile to adult phase in *Hedera helix*. — Pl. Syst. Evol., Suppl. 2, 105—110.

SCHMALENBERGER, B., NAGL, W.: Different DNA content, chromatin condensation and transcription activity in the retina of the Guinea-pig. — Pl. Syst. Evol., Suppl. 2, 119—125.

SCHWEIZER, D., NAGL, W., 1976: Heterochromatin diversity in *Cymbidium*, and its relationship to differential DNA replication. — Experim. Cell Res. 98, 411—423.

SLUYSER, M., 1977: The H 1 histones. — Trends Biochem. Sci. 2, 202—204.

SPURR, A. R., 1969: A low viscosity epoxy resin embedding medium for electron microscopy. — J. Ultrastr. Res. 26, 31—43.

TANAKA, R., 1971: Types of resting nuclei in *Orchidaceae*. — Bot. Mag. Tokyo 84, 118—122.

THOMA, F., KOLLER, TH., 1977: Influence of histone H 1 on chromatin structure. — Cell 12, 101—107.

TSCHERMAK-WOESS, E., 1963: Strukturtypen der Ruhekerne von Pflanzen und Tieren. — Wien: Springer.

WORCEL, A., BENYAJATI, C., 1977: Higher order coiling of DNA in chromatin. — Cell 12, 83—100.

YUNIS, J. J., YASMINEH, W. G., 1971: Heterochromatin, satellite DNA, and cell function. — Science 174, 1200—1209.

Address of the author: Prof. Dr. WALTER NAGL, Department of Biology, The University, P. O. Box 3049, D-6750 Kaiserslautern, Federal Republic of Germany.

Nucleotype and Heterochromatin

Pl. Syst. Evol., Suppl. 2, 263—280 (1979)
© by Springer-Verlag 1979

Institute of Botany, University of Vienna, Austria

Evolutionary Changes of DNA and Heterochromatin Amounts in the *Scilla bifolia* Group (*Liliaceae*)*

By

Johann Greilhuber, Wien

Key Words: *Liliaceae*, *Scilleae*, *Scilla bifolia group*, *Chionodoxa* BOISS.—DNA contents, heterochromatin contents; species diversification, chromosome evolution; cytosystematics.

Abstract: The amounts of nuclear DNA and constitutive heterochromatin have been determined in 18 species of the *Scilla bifolia* group, and in the less closely related *S. messeniaca* and *S. siberica*. Together with morphological criteria these data allow to reconstruct the pathways of chromosome evolution in the *S. bifolia* group at the diploid level with considerable accuracy. A progressive decrease of the DNA content by a factor of 0.5 occurred during the evolution from primitive yellow-seeded (*S. kladnii*: 1 C = 8.6 pg) to advanced black-seeded species (e.g. *S. luciliae*: 1 C = 4.3 pg). Grey- and brown-seeded species correspond to intermediary evolutionary stages and have intermediate DNA contents. 14 of 18 species have low amounts of heterochromatin (appr. 4 %) irrespective of DNA content. However, two separate evolutionary side-branches are characterized by the accumulation of C-band material (3 yellow-seeded and 1 black-seeded species). The data suggest that an incipient DNA increase due to heterochromatin addition occurred in both instances. Further speciation in the yellow-seeded branch was accompanied by a decrease of total DNA despite of an increase in C-band content.—DNA contents once more establish the close relationship of the *S. nivalis* subgroup with the taxa formerly classified under *Chionodoxa*, viz. the *S. luciliae* subgroup.— The systematic position of *S. messeniaca* (x = 9) next to the *S. bifolia* group, previously suggested for reasons of morphology and chromosome number, is now supported by the DNA amount (1 C = 10.7 pg). *S. siberica* (x = 6) as a remote relative contains considerably more DNA (1 C = 31.7 pg).

Apart from polyploidy, there are at least two further nuclear phenomena associated with evolutionary divergence among the taxa related to *Scilla bifolia*: changes in DNA content, and changes in amount

---

* Evolution of *Scilla* and Related Genera, VI.

and distribution of constitutive heterochromatin (Greilhuber & Speta 1977, Greilhuber 1978). While previous investigations were confined to the study of Giemsa banding and DNA contents in the Central European species *Scilla bifolia*, *S. drunensis*, and *S. vindobonensis*, the present study is concerned with the DNA and heterochromatin contents in the *S. bifolia* group on a larger scale. Two main questions have served as guide lines: 1) Is there a correlation of these nuclear characters with the taxonomic grouping on a morphological basis? 2) Can we gain some insight about the direction of evolution, and in particular of chromosome evolution, within this group of related species?

Materials and Methods

Species and Provenances. The taxa investigated are cultivated in the Botanical Garden of the City of Linz (Austria), or in the Botanical Garden of the University of Vienna. They have been collected in the natural habitat or were partly obtained from a commercial breeder. Herbarium specimens are preserved in the herbarium of F. Speta, Linz (Sp).

Scilla bifolia group:

1) *S. kladnii* Schur. (a) Rumania: Baile 1. Mai, 22. 3. 1974, F. Speta (Sp). (b) Hungary: woodland near Tarpa, 16. 3. 1974, F. Speta (Sp).

2) *S. decidua* Speta. Turkey: prov. Bolu, Bolu dağı pass, approx. 1000 m alt., 29. 3. 1974, F. Speta (Sp). Type material!

3) *S. vindobonensis* Speta. (a) Austria, Vienna: Meadows of the "Prater", City of Vienna, 160 m alt., 3. 4. 1977, J. Greilhuber. (b) Lower Austria: Haunoldstein im Pielachtal, district of St. Pölten, 235 m alt., 28. 3. 1973, F. Speta (Sp).

4) *S. voethorum* Speta. Greece: Isle of Samos, near Kokaria, 1976, W. Vöth (Sp). Type material!

5) *S. resslii* Speta. Turkey: prov. Ankara, Akyarma, 30 km NW Kızılcahamam, 1500 m alt., 11. 4. 1974, F. Ressl (Sp). Type material!

6) *S. bifolia* L. Germany, FRG: (a) Saarland, St. Gangolf/Saar (Sp). (b) Bavaria, near Fristingen, district Dillingen/Donau, appr. 420 m alt., 24. 4. 1971, 4. 1975, W. Sauer (Sp). (c) Upper Austria: Meadows "Pleschinger Au", Plesching, near Linz, F. Speta (Sp). (d) Upper Austria: Water meadows along the Danube near Aschach, 1.8 km downstream at Brandstatt, 266 m alt., 29. 3. 1972, F. Speta (Sp). (e) Lower Austria: Water meadows along the Danube near Stopfenreuth, 144 m alt., 7. 4. 1975, J. Greilhuber. (f) Italy, Umbria, prov. Perugia: Monti Reattini, Mte. Terminillo, 7. 6. 1977, E. Vitek.

7) *S. drunensis* (Speta) Speta. Upper Austria: Pucking, near Traun, "Oberreiter", 285 m alt., 20. 3. .1977, J. Greilhuber.

8) *S. buekkensis* Speta. Hungary: Bükk Mountains, 1.5 km NW Répáshuta, 23. 3. 1974, F. Speta (Sp). Type material!

9) *S. laxa* Schur. Rumania: Talmaciu at Sibiu, hill "Burgberg", 20. 3. 1974, F. Speta (Sp).

10) *S. pneumonanthe* Speta. Greece: Peloponnisos, mt. Chelmos, 2200 m alt., 13. 4. 1974, F. Speta (Sp). Type material!

11) *S. xanthandra* C. Koch (syn.: *S. uluensis* Speta).[1] Turkey: prov. Bursa, mt. Uladağ near Bursa, 1800 2000 m alt., 24. 4. 1972, W. Vöth (Sp).

---

[1] Synonymy according to F. Speta, pers. comm.

12) *S. nivalis* Boiss. Cultivar of the Botanical Garden of the City of Linz (Sp).

13) *S. siehei* (STAPF) SPETA. (a) cv. "Zwanenburg", obtained from TUBERGEN Ltd., Haarlem, The Netherlands, 1977. (b) cv. "Rosea", obtained from TUBERGEN Ltd. 1977.

14) *S. sardensis* (WHITTALL ex BARR et SUGDEN) SPETA. (a) Cultivar, obtained from TUBERGEN Ltd. 1976, 1977. (b) Cultivar of the Botanical Garden of the City of Linz.

15) *S. luciliae* (Boiss.) SPETA. Cultivar, obtained from TUBERGEN Ltd. 1977.

16) *S. nana* (J. A. et J. H. SCHULTES) SPETA. Greece, Crete (Kriti): Omalos plateau, above Xiloskalon, 6. 5. 1971, H. MALICKI (Sp).

17) *S. albescens* SPETA. Greece, Crete (Kriti): (a) Ida mts., peak-region, 2400 m alt., 9. 5. 1971, H. MALICKI (Sp). (b) Nomos Rethimnis, Ep. Milopotamu, Ida mts. (Psiloriti); western part of the Timpanato crest, N. Nida plateau, on the way to Anojia, 1500 m alt., 6. 4. 1971, G. & W. SAUER (Sp).

18) *S. tmoli* (WHITTAL) SPETA. Cultivar, obtained from TUBERGEN Ltd. 1977.

Scilla messeniaca group:
S. messeniaca Boiss. Seeds of cultivated plants were obtained from the Botanical Garden of the City of Triest, Italy.

Scilla siberica group:
S. siberica HAW. in ANDR. cv. "Spring Beauty" (triploid variety). Cultivated in the Botanical Garden of the City of Linz, originally obtained from TUBERGEN Ltd. 1. 4., 29. 4. 1973 (Sp).

Feulgen Staining and Cytophotometry. The two-wavelength method of PÁTAU (1952) was employed on a Leitz MPV 2 cytophotometer. The Feulgen schedule was the same as described previously (GREILHUBER 1977, 1978). *Scilla* species were measured against *Allium cepa* cv. "Ailsa Craig" (recommended internal standard, see BENNETT & SMITH 1976), or against other *Scilla* species. In the latter case the relative values were converted to absolute ones by using any available absolute value of an included *Scilla* species (measured against *A. cepa*) as separate calibration standard, and averaging the obtained amounts. A major methodical deviation from the previously used scheme was, that the background was not determined over an empty area but over the cell wall covering the nucleus to be measured. Both ways create errors which are discussed below. Histograms of isolated interphase nuclei were performed as described in GREILHUBER (1978).

Heterochromatin Quantification. The data are in part already published (GREILHUBER & SPETA 1977, 1978). In the remaining species the quantities have been estimated from C-band preparations as described previously. Details of C-banding will be presented in a forthcoming paper.

Results and Discussion

Infraspecific Variation and Methodical Errors. All questions about the taxonomical and evolutionary significance of DNA contents have to be preceded by considerations about the infraspecific variability and—first of all—about major sources of methodical errors which may influence the results of measurements and eventually mimic genuine differences in nuclear DNA amounts (see e.g. BENNETT & SMITH 1976). For the present investigation it was interesting how the background selection, over the cell wall or an empty area, influences the values, since the present procedure differed in this respect from a previous one applied in the same group (GREILHUBER 1978). Errors are immanent in both schedules: either one measures too much coagulated cytoplasm as background, or the cell walls together with some cytoplasm additional to the nucleus. A preliminary test (Table 1) leads one to expect a difference of about 10% in species like *S. bifolia* dependent on the background chosen. This fully explains why in the present investigation consistently somewhat lower values were measured for *S. bifolia* and *S. vindobonensis* than previously (GREILHUBER 1978), even in the same provenances (*S.*

Table 1. Systematic errors in measuring non-isolated Feulgen stained nuclei (telophase) dependent on background selection. The same nuclei have been measured setting as background (100% transmission) an empty area or the cell containing the respective nucleus

Series	Species	Slide no.	N	Background: Cell	Background: Empty area	Difference
1	*S. bifolia*	1	10	433 ± 11	486 ± 10	+ 12.4%
	(prov. 8 b)	2	10	448 ± 26	480 ± 34	+ 7.3%
	A. cepa cv.	1	10	1325 ± 62	1370 ± 58	+ 3.4%
	'Ailsa Craig'	2	10	1342 ± 31	1409 ± 21	+ 5.0%
2	*S. luciliae*	1	10	331 ± 11	353 ± 17	+ 6.5%
	A. cepa cv. 'Ailsa Craig'	1	10	1306 ± 50	1338 ± 48	+ 2.5%

bifolia: 34.1 versus 38.3% of *Allium cepa*; *S. vindobonensis*: 54.9 versus 59.0% of *A. cepa*).

The question of infraspecific variability of the DNA amount was investigated sufficiently extensively in *S. bifolia* (9 series of measurements, 6 distant provenances). The results (Table 6) clearly justify the conclusion that the DNA contents are stable within the resolution limits of this method.

DNA Contents, Heterochromatin Amounts, and Taxonomic Grouping. An about 2.2-fold variation of DNA amounts exists among the 18 investigated species related to *S. bifolia* s. str. (Tables 2-5). Furthermore, there are several species which differ from the majority of the taxa by their significantly larger heterochromatin content (Table 5). How do these nuclear characters relate to the taxonomical grouping on a morphological basis?

Table 2. DNA values in species of the *Scilla bifolia* group, *S. messeniaca*, and *S. siberica* as obtained from calibration against *Allium cepa* cv. 'Ailsa Craig' (Species, provenances, haploid chromosomes number n, pg DNA, DNA amount relative to *A. cepa*, number of nuclei measured, and the number of the series

Species	n	pg (1 C)	% of *A. cepa*	N	series no.
S. kladnii (3 a)	9	8.9 ± 0.3	52.9	50	1.
S. decidua	9	7.0 ± 0.4	41.9	46	2.
S. vindobonensis (5 a)	9	9.2 ± 0.4	54.9	30	3
S. voethorum	9	9.2 ± 0.3	54.7	55	4.
S. resslii	9	7.6 ± 0.3	45.1	50	5.
S. bifolia (8 a)	9	5.8 ± 0.3	34.6	30	6.
S. bifolia (8 b)	9	5.5 ± 0.3	33.0	21	7.
S. bifolia (8 b)	9	5.8 ± 0.3	34.8	20	8.
S. bifolia (8 e)	9	5.7 ± 0.2	33.8	29	9.
S. buekkensis	18	12.1 ± 0.7	72.3	52	10.
S. laxa	18	10.4 ± 0.4	62.3	50	3.
S. nivalis	9	3.9 ± 0.3	23.5	29	11.
S. xanthandra	9	4.0 ± 0.2	23.8	31	12.
S. pneumonanthe	27	16.3 ± 0.9	97.5	30	13.
S. sardensis	9	4.4 ± 0.2	26.2	30	14.
S. siehei cv. 'Zwanenburg'	9	4.2 ± 0.2	25.2	32	14.
S. siehei cv. 'rosea'	9	4.8 ± 0.3	28.9	31	10.
S. luciliae	9	4.2 ± 0.3	25.1	36	14.
S. messenica	9	10.7 ± 0.7	64.1	53	10.
S. siberica[1]	6	31.7	189.1	21	15.

[1] Value calculated for the diploid level, 2 n = 12, 2 x. 2 n = 18, 3 x: 47.5 ± 1.1 pg, 283.6% of *A. cepa*.

According to SPETA (1976 a) the *S. bifolia* group can be further divided best by the colour of the ripe seeds. He distinguishes a *S. vindobonensis* subgroup, in which he includes the yellow-seeded species *S. vindobonensis*, *S. voethorum*, *S. resslii*, *S. kladnii*, *S. taurica*, and the grey-seeded *S. decidua*, further a *S. bifolia* subgroup, which contains the brown-seeded species *S. bifolia*, *S. drunensis*, *S. buekkensis*, and *S. laxa*, and a *S. nivalis* subgroup, which embraces the black-seeded species *S. nivalis*, *S. xanthandra*, *S. longistylosa*, and *S. pneumonanthe*. Furthermore, a fourth subgroup is represented by the black-seeded *S. luciliae* subgroup, which was previously classified under *Chionodoxa* BOISS. It can be distinguished from the *S. nivalis* subgroup by a partly connate perigon and broader filaments. Still, the close relationship of the two subgroups is documented—apart from morphological criteria—by the

J. Greilhuber:

Table 3. DNA values in species of the *Scilla bifolia* group relative to each other and recalibrated against other *Scilla* species. Series no., species (provenance), relative DNA content ($\bar{x} \pm s$; %), 1C value (pg), number of nuclei (N) measured. Data for recalibration see Table 2

Series no.	Species	Relative DNA content (%)	1 C value (pg)	N
16.	*S. albescens* (19 b)	65.8 ± 3.7	3.7	30
	S. vindobonensis (5 b)	165.6 ± 6.2	9.4	30
	S. bifolia (8 c)	100.0 ± 3.1	5.7	27
17.	*S. voethorum*	161.4 ± 5.0	9.0	20
	S. resslii	137.2 ± 4.4	7.6	50
	S. vindobonensis (5 b)	172.0 ± 5.1	9.6	50
	S. kladnii (3 b)	158.4 ± 3.5	8.8	50
	S. decidua	128.2 ± 4.3	7.1	20
	S. bifolia (8 d)	100.0 ± 5.5	5.6	20
18.	*S. xanthandra*	80.0 ± 4.9	4.6	39
	S. pneumonanthe	269.7 ± 13.4	15.4	50
	S. kladnii (3 a)	144.0 ± 5.2	8.2	50
	S. bifolia (8 e)	100.0 ± 3.9	5.7	50
19.	*S. sardensis*	67.0 ± 4.4	3.9	47
	S. nivalis	70.5 ± 3.2	4.1	49
	S. bifolia (8 e)	100.0 ± 4.1	5.8	52
20.	*S. nivalis*	107.5 ± 8.0	4.3	50
	S. nana	105.9 ± 4.2	4.3	45
	S. albescens (19 a)	94.2 ± 4.6	3.8	49
	S. sardensis	100.0 ± 6.2	4.0	58

formation of fertile hybrids between some species of the *S. luciliae* subgroup and *S. nivalis* (see Speta 1976 a).

The DNA and heterochromatin amounts are diagrammatically presented in Fig. 1 and confronted with the taxonomic grouping according to Speta (1976 a, b, and pers. comm.), the seed colour and the perigon shape as the most relevant morphological traits, and—as a further important character—the embryosac type.[1] It is recognized that the higher DNA values per basic genome occur in the *S. vindobonensis* subgroup (9.4–7.1 g), medium values in the *S. bifolia* subgroup (6.0–5.2 pg), while low values (5.3–3.8 pg) occur both in the *S. nivalis* and *S. luciliae* subgroup. The grouping based on seed colour is therefore fully compatible with the distribution of DNA amounts, and

---

[1] The embryological data have been worked out by Erika Svoma and will be published by her in detail in the near future.

lends support to the conclusion that both traits reflect natural relationships.

However, there are still more or less conspicuous differences within the subgroups:

S. vindobonensis **Subgroup.** 5 species were investigated; *S. taurica* was not available. *S. kladnii* and *S. decidua* are poor in heterochromatin and have monosporic embryosacs, but are distinguished by the seed colour and the absolute DNA amount, which is lower in *S. decidua*. *S. vindobonensis*, *S. voethorum* and *S. resslii* are rich in heterochromatin, and are characterized by a tetrasporic embryosac. It may be added that the Giemsa banding patterns in these species also point to their close relationship (unpublished observations). DNA content and heterochromatin amount are inversely correlated in these species.

S. bifolia **Subgroup.** Three of the four presently recognized species are tetraploids (higher levels of polyploidy and aneuploids occur occasionally in *S. buekkensis* and *S. laxa*, see also SPETA 1976 b). They all show a low amount of heterochromatin, and a lower level of DNA content than the *S. vindobonensis* subgroup. While the basic genomes in *S. bifolia*, *S. drunensis* and *S. buekkensis* do not seem to be significantly different, the genome of *S. laxa* appears to be slightly smaller than that of *S. bifolia* (5.2 versus 5.7 pg). The significance of this difference will have to be established in forthcoming investigations. All taxa have monosporic embryosacs.

S. nivalis **Subgroup.** 4 species are recognized, viz. *S. nivalis*, *S. xanthandra*, *S. longistylosa*, and *S. pneumonanthe; S. longistylosa* was not available. All species are poor in heterochromatin. The DNA amounts of the diploid species *S. nivalis* and *S. xanthandra* (4.1 and 4.3 pg respectively) are clearly below the level in *S. bifolia* and obviously not significantly different from each other. However, the hexaploid *S. pneumonanthe* (5.3 pg) shows a comparatively high DNA content per basic genome. The embryosacs are monosporic in this subgroup.

S. luciliae **Subgroup.** 8 species are recognized by SPETA (1976 a), 6 of which are included in this investigation. *S. tmoli* is the only among these which is rich in heterochromatin. A further distinction can be made on the basis of the embryosac type: *S. sardensis* and *S. siehei* have monosporic embryosacs, but *S. luciliae*, *S. nana*, *S. albescens*, and *S. tmoli* have tetrasporic ones. Practically identical DNA amounts are found in *S. sardensis*, *S. luciliae*, and *S. nana*. The slightly higher value in *S. siehei* may not be genuine. Likewise, the slightly lower, DNA content in *S. albescens*, as compared with the closely related *S. nana*, is within the range of methodical variation. On the basis of the presumably more reliable histograms of isolated interphase nuclei the slightly higher DNA content calculated for *S. tmoli* (about 8 %, i.e. 0.4–0.5 pg above the level of the other related species: Fig. 2) may be genuine. This amount of excess DNA in *S. tmoli* is in quantitative agreement with the approx. 0.4 pg surplus heterochromatin in this species. We can assume therefore that the amount of euchromatic DNA is fairly constant within the *S. luciliae* subgroup.

Table 4. Histogram comparison (G 1 peaks obtained from isolated interphase nuclei) of *S. bifolia*, *S. sardensis*, *S. luciliae*, and *S. tmoli*. Absolute values used for recalibration see Table 2

Series no.	Species	Relative DNA content (%)	1 C value (pg)	N
21.	*S. tmoli*	82.7 ± 3.7	4.7	147
	S. sardensis	77.1 ± 3.8	4.4	128
	S. luciliae	76.0 ± 3.6	4.3	97
	S. bifolia (8 f)	100.0 ± 4.5	5.7	71

Distantly Related Taxa: *S. messeniaca*, *S. siberica*, and *S. persica*. Among the species presently classified under *Scilla* L. only the species of the *S. siberica* group and *S. messeniaca* represent taxonomic groups which are undoubtedly, though ± distantly, related with the *S. bifolia* group. In particular *S. messeniaca* is considered to show clear affinities to *S. bifolia* s.l. Respective arguments are the same chromosome number (x = 9), a similar chromosome size, and the similarities in bracts, flowers and seeds. However, essential differences are found in bulb shape, colour of the tunica, leaf number (3–7), scape cross section, inflorescence, and ontogenetic origin of the elaiosome (SPETA 1974). It was therefore of interest to include in the present study the DNA amounts of *S. messeniaca* as the closest, of *S. siberica* as a remote relative of the *S. bifolia* group, and of *S. persica* as a particularly primitive species within *Scilla* L. Among these taxa *S. persica* (1 C = 21.0 pg, n = 4, see GREILHUBER 1977) and *S. siberica* (1 C = 31.7 pg, n = 6) show comparatively high DNA amounts. *S. messeniaca* with 1 C = 10.7 pg, n = 9, however, is not much above the upper limit in the *S. bifolia* group (1 C = 9.4 pg in *S. vindobonensis*). Therefore, the DNA contents too argue in favour of a closer taxonomic affinity of *S. messeniaca* and the *S. bifolia* group.

Table 5. Survey of DNA amounts (average values, calculated for the basic chromosome set, x) and heterochromatin amounts in the *Scilla* species discussed

Species	Chromosome number (n)	1 C value (pg) (x-level)	Hetero-chromatin amount (%)
S. persica	4	21.0	1.8[1]
S. siberica	6	31.7	19.9[2]
S. messeniaca	9	10.7	2.0
S. kladnii	9	8.6	~4.0
S. decidua	9	7.1	~4.0
S. vindobonensis	9	9.4	10.6[3]
S. voethorum	9	9.1	15.4
S. resslii	9	7.6	22.1
S. bifolia	9	5.7	~4.0[3]
S. drunensis	18	5.9	~4.0[3]
S. buekkensis	18	6.0	~4.0
S. laxa	18	5.2	~4.0
S. nivalis	9	4.1	~4.0
S. xanthandra	9	4.3	~4.0
S. pneumonanthe	27	5.3	~4.0
S. sardensis	9	4.2	~4.0
S. siehei	9	4.5	~4.0
S. luciliae	9	4.3	~4.0
S. nana	9	4.3	~4.0
S. albescens	9	3.8	~4.0
S. tmoli	9	4.7	13.5

[1] See GREILHUBER & SPETA 1976.
[2] See GREILHUBER & SPETA 1978.
[3] See GREILHUBER & SPETA 1977.

The Direction of Evolutionary Change. DNA amounts, heterochromatin contents, degrees of polyploidy, together with the main morphological and embryological traits (Fig. 1) reveal a remarkable diversity even within each one of the four taxonomic subgroups of the *S. bifolia* group. However, there are three characters which can be found associated in species of all four subgroups: a diploid chromosome number, a heterochromatin-poor karyotype, and a monosporic embryosac. The corresponding taxa are *S. kladnii*, *S. decidua*, *S. bifolia*, *S. xanthandra*, *S. nivalis*, *S. sardensis*, and *S. siehei*. As already noted there is an obvious correlation of DNA content and seed colour. It seems reasonably clear that these so-called "main branch" taxa can be arranged in a series according to seed colour and DNA content, and that this reflects a phylogenetic sequence. Even the direction of evolutionary change within this series can be determined with considerable certainty.

In respect of *S. messeniaca*, the only close relative of the *S. bifolia* group, we may regard yellow seeds and a high DNA amount as primitive characters. Among the investigated main branch taxa only *S. kladnii* fulfills these requirements. Furthermore, the exceptionally high frequency of more than two, up to four, leaves in this species (SPETA

Table. 6. Constancy of the DNA amount in *Scilla bifolia* s. str. DNA values as calibrated against *Allium cepa* (*) or *Scilla* species other than *S. bifolia*. Provenance, series of measurement, 1 C value, deviation from mean, number of nuclei measured (N), and series no. (Tables 2-4) from where the values for recalibration were taken

Provenance	Series	1 C value (pg)	Deviation from mean (%)	N	Standard spp. in series
8 a	6.	5.8*	+ 1.8	30	
8 b	7.	5.5*	— 3.0	21	
8 b	8.	5.8*	+ 2.3	20	
8 e	9.	5.7*	— 0.7	29	
8 c	16.	5.6	— 2.4	27	3.
8 d	17.	5.5	— 3.2	20	1., 2., 3., 4., 5.
8 e	18.	5.7	+ 0.5	50	1., 12., 13.
8 e	19.	6.1	+ 6.5	52	11., 14.
8 f	20.	5.6	— 1.6	71	14.
mean:		5.7			

1976 b) is certainly a primitive trait of systematic significance. On the other hand, the strongly specialized new acquisitions in flower morphology clearly identify the *S. luciliae* subgroup as the most advanced subgroup. Among the main branch taxa *S. sardensis* and *S. siehei*, the latter appears to be more progressive by its larger flowers and a showy white eye which lacks in *S. sardensis*.

Seed colour, the DNA amounts as well as the capacity for fertile interbreeding with *S. nivalis* indicate that the *S. luciliae* subgroup evolved directly from the *S. nivalis* subgroup. *S. nivalis* and *S. sardensis* show the most close relationship.

The grey-seeded *S. decidua* as an "advanced" member of the *S. vindobonensis* group and the brown seeded *S. bifolia* obviously assume an intermediary phylogenetic position between the yellow- and black-seeded species. Does the grey- or brown-seeded condition represent a transitory stage from the yellow- to the black-seeded condition? Or emerged grey, brown- and black-seeded species directly from one or a few yellow-seeded parent species? At present this question must remain unanswered, but an answer possibly could come from the discovery of further species.

In view of the main branch taxa, evolutionary side branches are easily recognized: 1) *S. vindobonensis*, *S. voethorum* and *S. resslii* of the *S. vindobonensis* subgroup have tetrasporic embryosacs and considerably more heterochromatin than *S. kladnii*, their only next of kin among the more primitive species. There is but a slight difference in DNA content between *S. kladnii* ($1 C = 8.6$ pg) and *S. vindobonensis* ($1 C = 9.4$ pg), which can be fully explained by an increase in heterochromatin in the latter. Consequently it has to be assumed that further heterochromatin increase occurred in *S. voethorum* and *S. resslii*, nevertheless accompanied by a parallel net decrease of total DNA, and evidently at the expense of euchromatic DNA. 2) A tetrasporic embryosac has certainly originated independently from the *S. vindobonensis* group a second time within the *S. luciliae* group. It is found in *S. luciliae*, *S. nana*, *S. albescens*, and *S. tmoli*. 3) *S. tmoli* is the only among these species which has developed a heterochromatin-rich karyotype; its fruit shape and seed size also point towards a derived systematic position (SPETA 1976 a). 4) Polyploids represent a separate category of evolutionary side branches and are not included in Fig. 3. Among the four polyploids measured, "unexpected" DNA amounts were found in *S. laxa*, where the basic DNA content ($1 C_x = 5.2$ pg) appears to be slightly smaller than expected for a relative of *S. bifolia* ($1 C = 5.7$ pg), and in *S. pneumonanthe* ($1 C_x = 5.3$ pg), where the DNA amount is considerably above the values found in the related diploids *S. nivalis* and *S. xanthandra* ($1 C = $ approx. 4.2 pg). These differences may have been determined be the parental species (as allopolyploidy can be expected), or they may have been obtained during subsequent diploidization. Both possibilities will have to be considered.

Conclusions. As demonstrated above the evolutionary pathways in the *S. bifolia* group can be reconstructed now with reasonable accuracy. The results are diagrammatically presented for the diploid level in Fig. 3.

There are two aspects which are important from the viewpoint of genome evolution: 1) a straight decrease of euchromatic DNA content, 2) an increase in heterochromatin content, restricted, however, to two separate evolutionary side branches. No strict correlation between changes of heterochromatin and DNA content can be recognized, since heterochromatin-rich genomes evolved at high as well as at low C-values, and since the majority of species retained a low heterochromatin content irrespective of DNA decrease. A negative correlation is observed within the sequence *S. vindobonensis* -*S. voethorum* -*S. resslii*. Decrease in euchromatin and increase in heterochromatin may be functionally connected in this subgroup. However, in view of the general independence of these two phenomena in the other taxa we may interpret this correlation more properly as a mere coincidence of two independent trends in genome evolution: The euchromatic DNA becomes reduced during evolution of new species in the *S. bifolia* group, even if heterochromatic DNA is progressively accumulated.

The values of eu- and heterochromatin in *S. vindobonensis* and *S. tmoli* support the view that heterochromatin is an additive component in genome evolution; i.e., its accumulation results in an equal increase in total DNA amount, at least incipiently. The additive nature of heterochromatin was demonstrated at the infraspecific level just recently in the *Scilla siberica* group (GREILHUBER & SPETA 1978). The same was shown for certain variable C-bands in man (GERAEDTS & al. 1975, SUMNER 1977). Saltatory replication (BRITTEN & KOHNE 1968) or inequal exchanges in satellite DNA containing chromosome segments are conceivable mechanisms of quantitative variation in constitutive heterochromatin, which would leave the euchromatic part of the genome quantitatively unaltered. However, mechanisms of accumulation of satellite DNA containing constitutive heterochromatin may exist (see SMITH 1976) which work at the true expense of euchromatic DNA, as suggested by the results of SINGH & al. (1976) on sex chromosome evolution in snakes.

It may be anticipated that the karyotype structure in the *S. bifolia* group remained largely constant despite a progressive reduction of DNA contents, at least in the heterochromatin-poor species (GREILHUBER & SPETA, in prep.). But even the existing differences between heterochromatin-rich and -poor species, as in *S. vindobonensis* and *S. bifolia*, may be largely due to conspicuous C-bands (GREILHUBER & SPETA 1977). Therefore highly dispersed changes in euchromatin are much more likely than major localized deletions. Highly and intermediately repetitive DNA fractions are preferred candidates for disproportionate reduction during evolution in the *S. bifolia* group, but unique sequences may have

---

Fig. 1.   Distribution of DNA contents ($1 C_x$—values) and heterochromatin amounts in species of the *Scilla bifolia* group, and *S. persica*, *S. siberica*, and *S. messeniaca* as more or less remote relatives. These nuclear characters are confronted with the seed colour, gross perigon morphology, the embryosac type, and the taxonomic subgrouping of the *S. bifolia* group after SPETA 1976 a (unpublished embryological data personally communicated by ERIKA SVOMA. Vienna)

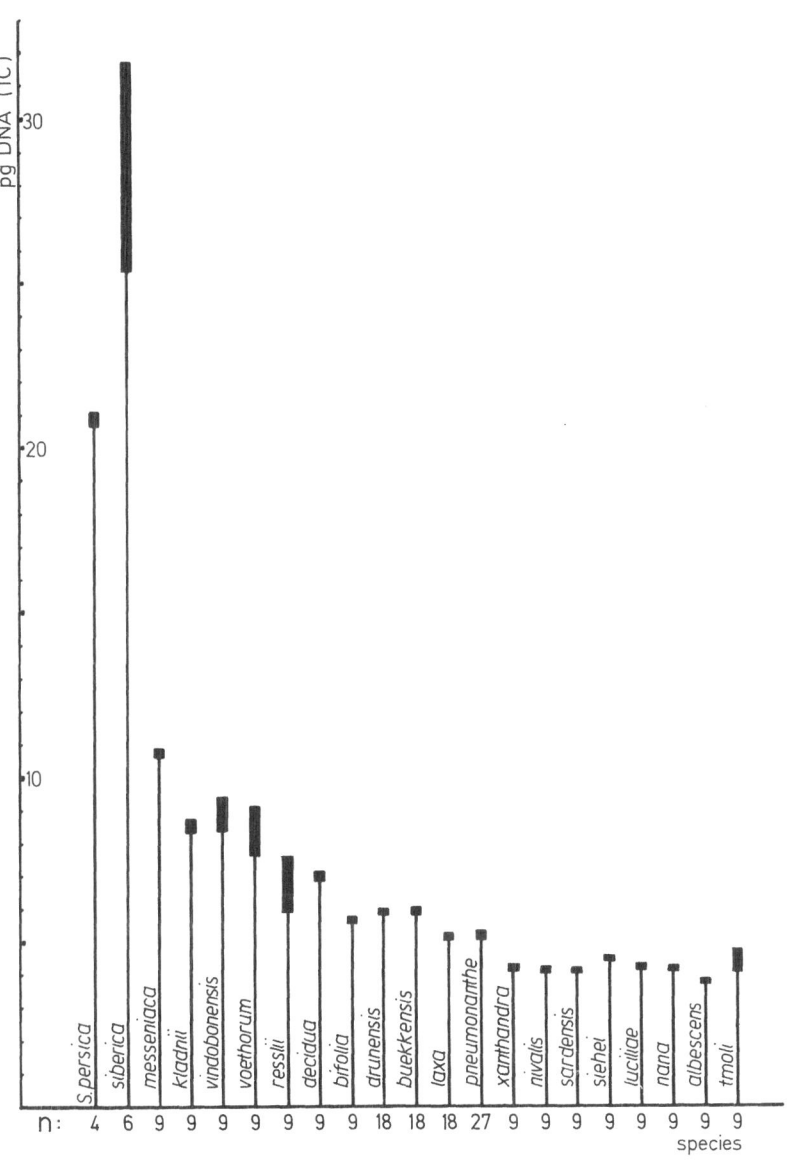

varied as well. The overall correlation of genome size and relative amount of repetitive DNA is generally acknowledged (Flavell & al. 1974). But at the level of more closely related taxonomic groups it becomes increasingly clear that the different DNA classes with fast,

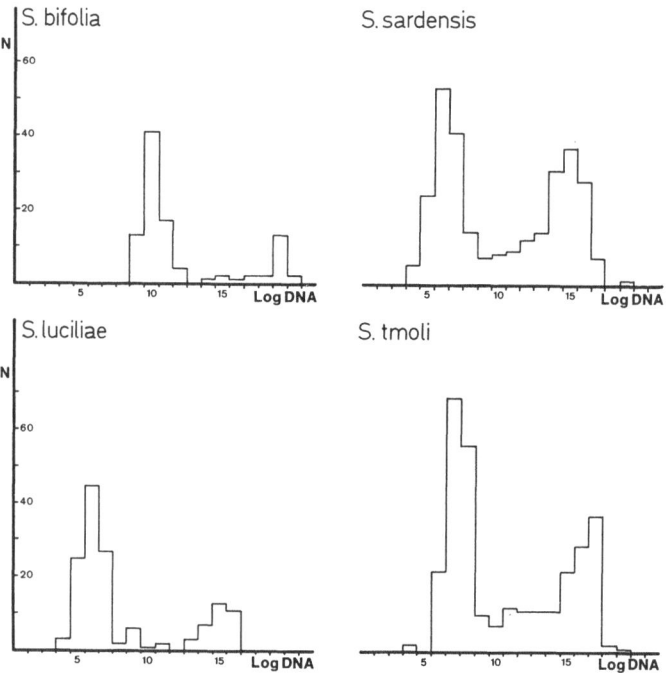

Fig. 2. Comparative histograms of isolated interphase nuclei in *S. bifolia*, *S. sardensis*, *S. luciliae*, and *S. tmoli* (compare Table 4). Abscissa: Log. DNA-Feulgen/nucleus, 10 size classes between 2 C and 4 C. Ordinate: no. of nuclei measured

intermediate and slow reassociation can be involved to quite different degrees in the increase or reduction of the genome sizes (e.g. Cullis & Schweizer 1973, Narayan & Rees 1976, Bachmann & Price 1977).

The functional significance of the nuclear changes in the evolution of the *S. bifolia* group is still not known. At present, the "nucleotypic" effects of DNA variation (Bennett 1972) lend themselves best to evolutionary interpretations, although remarkable differences in generation time are not found in these plants: 1) Cell cycle time varies directly proportional to DNA content, so that quicker growth can be

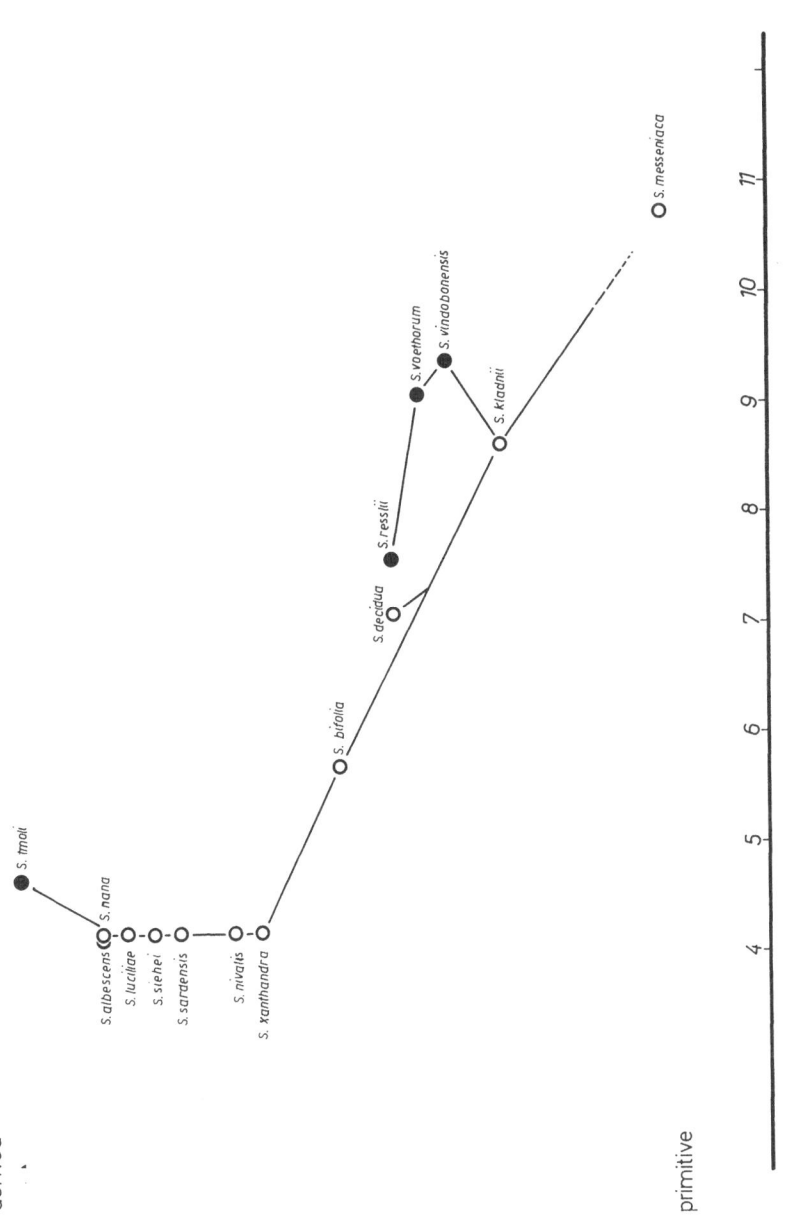

Fig. 3. An evolutionary interpretation of the interspecific variation of DNA and heterochromatin contents in the *Scilla bifolia* group. A clear trend from high to low DNA amounts can be recognized. A DNA increase seems to have occurred only twice; it is likely that it was solely due to accumulation of constitutive heterochromatin (solid circles: heterochromatin-rich species)

accomplished when the genome size becomes reduced. 2) Cell size varies directly proportional to DNA content (PRICE & al. 1973); the developmental effects of cell cycle time may thus be counteracted or even overruled (PRICE & BACHMANN 1976). 3) There is evidence for disproportionately short cell cycles in heterochromatin-rich species (NAGL 1974, NAGL & EHRENDORFER 1974, BÖSEN & NAGL 1978); heterochromatin, therefore, may be a compensatory mechanism for the cell cycle effects of evolutionary DNA increase. The relative importance of these nucleotypic effects in the *S. bifolia* group and their relation to the environment is largely accessible to investigation. Furthermore, and apart from nucleotypic effects of this type, there is the distinct possibility that heterochromatin, which is known to alter meiotic recombination drastically (e.g. MIKLOS & NANKIVELL 1976), has played a significant role in the evolution of the heterochromatin-rich species in the *S. bifolia* group by specifically transforming their genetic system.

Summary

1) Nuclear DNA contents (presented as 1 C values) have been determined by two-wavelength Feulgen-cytophotometry in 17 species of the *S. bifolia* group, and in *S. messeniaca* and *S. siberica* as more or less related taxa (survey see Table 5, p. 271). C-values are now available in 18 of the 22 species presently acknowledged in the *S. bifolia* group. The amount of constitutive heterochromatin has been estimated from C-banded metaphase chromosomes.

2) In the *S. bifolia* group C-values range from 9.4 to approx. 4.2 pg at the x-level. Their distribution is consistent with the supraspecific taxonomic grouping as proposed by SPETA (1976 a)—*S. vindobonensis* subgroup: seeds yellow or grey, high C-values; *S. bifolia* subgroup: seeds brown, intermediate C-values; *S. nivalis* subgroup: seeds black, low C-values; *S. luciliae* group: seeds also black, but perigon partly connate, low C-values.

3) Nuclear together with morphological criteria allow to reconstruct evolutionary pathways, and to describe chromosome evolution in terms of changes in DNA and heterochromatin content. A progressive decrease to half of the original DNA content occurred during the evolution from primitive yellow-seeded (*S. kladnii*: 1 C = 8.6 pg) to advanced black-seeded species (approx. 4.2 pg in the *S. nivalis* and *S. luciliae* group). Grey- (*S. decidua:* 1 C = 7.1 pg) and brown-seeded species (*S. bifolia:* 1 C = 5.7 pg) correspond to intermediary evolutionary stages with intermediate DNA contents.

4) The DNA contents once more establish the close relationship of the black seeded *S. nivalis* subgroup with the black-seeded, evolutionary

most advanced taxa formerly classified under *Chionodoxa* BOISS, viz. the *S. luciliae* subgroup.

5) 14 of 18 species in the *S. bifolia* group have low amounts (approx. 4%) of heterochromatin, independently of their C-value. However, accumulation of C-banding heterochromatin characterizes two separate evolutionary side-branches which have originated at high (3 yellow-seeded species) and low DNA levels (1 black-seeded species). The data suggest that in both instances the increase of heterochromatin quantity caused an incipient DNA increase. C-bands are therefore regarded as an additive component in genome evolution.

6) *S. messeniaca* is regarded as isolated monotypic group, but the only in the genus *Scilla* to be placed near the *S. bifolia* group (SPETA 1974). The DNA content (1 C = 10.7 pg) supports this view. *S. siberica* as a more remote relative contains considerably more DNA (1 C = 31.7 pg).

7) DNA contents were checked on a larger geographic scale within *S. bifolia*. There is no evidence for DNA variability beyond the level of the resolution of the method. Systematic errors in measuring non-isolated Feulgen-stained nuclei were found to account fully for the slight differences between the present and previous (GREILHUBER 1978) values in *S. bifolia* and *S. vindobonensis*.

I wish to thank Dr. F. SPETA for his permanent cooperation, particularly in regard to plant material and taxonomy. Special thanks go to Miss ERIKA SVOMA for her permission to quote some of her unpublished embryological results. The efforts of Mr. F. TOD, H. MUER, and W. VÖTH (Botanical Garden of the University of Vienna) in plant cultivation are gratefully acknowledged.—This investigation was financially supported by the "Hochschuljubiläumsstiftung der Stadt Wien" (Leitz MPV 2 cytophotometer).

References

BACHMANN, K., PRICE, H. J., 1977: Repetitive DNA in *Cichorieae* (*Compositae*). — Chromosoma (Berl.) **61**, 267—275.

BENNETT, M. D., 1972: Nuclear DNA content and minimum generation time in herbaceous plants. — Proc. R. Soc. Lond. B **181**, 109—135.

— SMITH, J. B., 1976: Nuclear DNA amounts in angiosperms. — Phil. Trans. R. Soc. Lond. B **274**, 227—274.

BÖSEN, H., NAGL, W., 1978: Short duration of the mitotic and endomitotic cell cycle in the heterochromatin-rich monocot *Allium carinatum*. — Cell Biol. Int. Rep. **2**, 565—571.

BRITTEN, R. J., KOHNE, D. E., 1968: Repeated sequences in DNA. — Science **161**, 529—540.

CULLIS, C. A., SCHWEIZER, D., 1974: Repetitious DNA in some *Anemone* species. — Chromosoma (Berl.) **44**, 417—421.

GERAEDTS, J. P. M., PEARSON, P. L., PLOEG, M., VAN DER, VOSSEPOEL, A. M., 1975: Polymorphism for human chromosomes 1 and Y: Feulgen and UV DNA measurements. — Exp. Cell Res. **95**, 9—14.

GREILHUBER, J., 1977: Nuclear DNA and heterochromatin contents in the *Scilla hohenackeri* group, *S. persica*, and *Puschkinia scilloides* (*Liliaceae*). — Pl. Syst. Evol. **128**, 243—257.

— 1978: DNA contents, Giemsa banding, and systematics in *Scilla bifolia*, *S. drunensis*, and *S. vindobonensis* (*Liliaceae*). — Pl. Syst. Evol. **130**, 223—233.

— SPETA, F., 1977: Giemsa karyotypes and their evolutionary significance in *Scilla bifolia*, *S. drunensis*, and *S. vindobonensis* (*Liliaceae*). — Pl. Syst. Evol. **127**, 171—190.

— — 1978: Quantitative analyses of C-banded karyotypes, and systematics in the cultivated species of the *Scilla siberica* group (*Liliaceae*). — Pl. Syst. Evol. **129**, 63—109.

MIKLOS, G. L., NANKIVELL, R. N., 1976: Telomeric satellite DNA functions in regulating recombination. — Chromosoma (Berl.) **56**, 143—167.

NAGL, W., 1974: Mitotic cycle time in perennial and annual plants with various amounts of DNA and heterochromatin. — Devel. Biol. **39**, 342—346.

— EHRENDORFER, F., 1974: DNA content, heterochromatin, mitotic index, and growth in perennial and annual *Anthemideae* (*Asteraceae*). — Pl. Syst. Evol. **123**, 35—54.

NARAYAN, R. K. J., REES, H., 1976: Nuclear DNA variation in *Lathyrus*. — Chromosoma (Berl.) **54**, 141—154.

PÁTAU, K., 1952: Absorption microspectrophotometry in irregularly shaped objects. — Chromosoma **5**, 341—362.

PRICE, H. J., BACHMANN, K., 1976: Mitotic cycle time and DNA content in annual and perennial *Microseridinae* (*Compositae*, *Cichorieae*). — Pl. Syst. Evol. **126**, 323—330.

— SPARROW, A. H., NAUMAN, A. F., 1973: Correlations between nuclear volume, cell volume, and DNA content in meristematic cells of herbaceous angiosperms. — Experientia **29**, 1028—1029.

SINGH, L., PURDOM, I. F., JONES, K. W., 1976: Satellite DNA and evolution of sex chromosomes. — Chromosoma (Berl.) **59**, 43—62.

SMITH, G. P., 1976: Evolution of repeated DNA sequences by unequal crossover. — Science **191**, 528—535.

SPETA, F., 1974: *Scilla messeniaca* BOISS. (*Liliaceae*) und ihre verwandtschaftlichen Beziehungen. — Ann. Mus. Goulandris **2**, 59—67.

— 1976 a: Über *Chionodoxa* BOISS., ihre Gliederung und Zugehörigkeit zu *Scilla* L. — Naturk. Jahrb. Stadt Linz **21** (1975), 9—79.

— 1976 b: Cytotaxomischer Beitrag zur Kenntnis der *Scilla*-Arten Ungarns und Siebenbürgens. — Naturk. Jahrb. Stadt Linz **22** (1976), 9—63.

SUMNER, A. T., 1977: Estimation if the sizes of polymorphic C-bands in man by measurement of DNA content of whole chromosomes. — Cytogenet. Cell Genet. **19**, 250—255.

Address of the author: Univ.-Doz. Dr. JOHANN GREILHUBER, Institut für Botanik der Universität Wien, Rennweg 14, A-1030 Wien, Austria.

Index to Methods, Subjects, and Taxa